Selected Mathematical Methods in Theoretical Physics

T0187564

Selected Mathematical Methods in Theoretical Physics

Vladimir P. Krainov
Moscow Institute of Physics and Technology, Moscow, Russia

CRC Press
Taylor & Francis Group
Boca Raton London New York

CRC Press is an imprint of the
Taylor & Francis Group, an **informa** business

A TAYLOR & FRANCIS BOOK

First published 2002 by Taylor & Francis

Published 2019 by CRC Press
Taylor & Francis Group
6000 Broken Sound Parkway NW, Suite 300
Boca Raton, FL 33487-2742

© 2002 by Taylor & Francis Group, LLC
CRC Press is an imprint of Taylor & Francis Group, an Informa business

No claim to original U.S. Government works

ISBN 13: 978-0-415-27239-1 (pbk)

Visit the Taylor & Francis Web site at
http://www.taylorandfrancis.com

and the CRC Press Web site at
http://www.crcpress.com

This book has been produced from camera-ready copy supplied by the authors

Every effort has been made to ensure that the advice and information in this book is true and accurate at the time of going to press. However, neither the publisher nor the authors can accept any legal responsibility or liability for any errors or omissions that may be made. In the case of drug administration, any medical procedure or the use of technical equipment mentioned within this book, you are strongly advised to consult the manufacturer's guidelines.

British Library Cataloguing in Publication Data
A catalogue record for this book is available from the British Library

Library of Congress Cataloging in Publication Data
A catalogue record has been requested

CONTENTS

Chapter 1

Calculation of Integrals

1.1 SADDLE-POINT METHOD

Elementary functions, for example, $f(x) = \cos x$ are well known to students. Higher transcendental functions can be represented as a rule in the form of integrals from certain elementary functions, or from simpler transcendental functions. The goal of this chapter is to simplify a typical higher transcendental function in various regions of its argument and of its index and express it via elementary functions.

We choose *the Bessel function* $J_v(x)$ as such an example. This function is expressed in the form of the definite integral

$$J_v(x) = \frac{1}{2\pi} \int_{-\pi}^{\pi} \exp[i(x \sin t - vt)] dt. \qquad (1.1)$$

If the index v is noninteger, then Eq. (1.1) determines the so-called *Anger function*.

In order to calculate integral (1.1), we suggest that the index $v \gg 1$. It should be noted that the results obtained below have high numerical accuracy, comparable to values of this index of the order of 1. Indeed, for example, the factorial $v!$ is described by the asymptotic Stirling formula even at $v = 1$ with a high accuracy of 8%. It should be noted that analytical derivation is possible also in the case of large arguments $x \gg 1$ and arbitrary values of the index v.

In such cases we can use the so-called *saddle-point method*. According to this method we determine the region of the integrand in the integral

1

$$I = \int \exp[f(t)]dt \,, \tag{1.2}$$

where the function $f(t)$ has a sharp maximum or it oscillates very slowly. We simplify the function $f(t)$ in this region, i.e. in the vicinity of the argument $t = \alpha$, which is determined from the equation $f'(\alpha) = 0$. Then we can calculate analytically the integral in Eq. (1.2). The exponent in the integrand of Eq. (1.2) is expanded in a Taylor series:

$$f(t) = f(\alpha) + \frac{1}{2} f''(\alpha)(t - \alpha)^2 \,.$$

Then using the value of the well-known *Poisson integral*

$$\int_{-\infty}^{\infty} \exp(-x^2)dx = \sqrt{\pi},$$

we obtain finally from Eq. (1.2):

$$I = \sqrt{\frac{2\pi}{-f''(\alpha)}} \; \exp[f(\alpha)]. \tag{1.3}$$

This procedure will be used for the Bessel function, Eq. (1.1) in various regions of its argument x. For the sakes of simplicity we will restrict ourselves to positive values of the argument x only.

1.2 THE REGION $x > v \gg 1$

Let us introduce the notation

$$f(t) = i(x \sin t - vt).$$

The extremum of this function is determined from the equation

$$f'(t = \alpha) = 0.$$

Thus, we obtain the equation $\cos\alpha = v/x$, i.e. the value of α is real at $x >$ v. In the integration interval $[-\pi,\ \pi]$ we have two values of α. Hence, it is useful to present the integral in Eq. (1.1) as a sum of two mutually complex conjugate expressions

$$J_\nu(x) = \frac{1}{2\pi} \int_0^\pi \exp[i(x \sin t - \nu t)]dt + \text{c.c.} \qquad (1.4)$$

The second derivative is of the simple form

$$f''(\alpha) = -ix \sin \alpha.$$

According to the general result (1.3) of the saddle-point method we obtain the approximate expression for the Bessel function

$$J_\nu(x) = \frac{1}{2\pi} \sqrt{\frac{2\pi}{ix \sin \alpha}}\ \exp[i(x \sin \alpha - \nu\alpha)] + \text{c.c.}$$

or

$$J_\nu(x) = \sqrt{\frac{2}{\pi x \sin \alpha}}\ \cos\left(x \sin \alpha - \nu\alpha - \frac{\pi}{4}\right) \qquad (1.5)$$

Since $x = v / \cos \alpha$, we find the relation $x \sin \alpha = v \tan \alpha$. Finally the so-called *Debye expansion* of the Bessel function at $x > v \gg 1$ follows from Eq. (1.5):

$$J_\nu\left(\frac{v}{\cos \alpha}\right) = \sqrt{\frac{2}{\pi v \tan \alpha}}\ \cos\left(v \tan \alpha - \nu\alpha - \frac{\pi}{4}\right) \qquad (1.6)$$

1.3 THE REGION $x \gg v$

We can simplify Eq. (1.5) by putting $\alpha \to \pi / 2$:

$$J_v(x) = \sqrt{\frac{2}{\pi x}} \cos\left(x - \frac{\pi v}{2} - \frac{\pi}{4} \right) \tag{1.7}$$

This asymptotic representation of the Bessel function for large values of its argument $x \gg 1$ is well known. It is correct not only for large values of the index $v \gg 1$, but also for $v < 1$ in particular, for $v = 0$. This result follows from the fact that in this case the saddle-point method is applicable due to large values of the argument $x \gg 1$ only, independent of the value of the index v.

In particular, the asymptotic representation (1.7) is valid also for the Bessel function $J_0(x)$ at $x \gg 1$.

1.4 THE REGION $x - v \ll v$

If the argument x nearly coincides with the index v (but not too closely, see below Sect. 1.8) and $x > v$, then we can simplify Eq. (1.6) using relations

$$\alpha < 1, \qquad \tan \alpha \approx \alpha + \frac{1}{3}\alpha^3 .$$

Hence,

$$J_v\left(v + \frac{1}{2}v\alpha^2 \right) = \sqrt{\frac{2}{\pi v\alpha}} \cos\left(\frac{1}{3}v\alpha^3 - \frac{\pi}{4} \right) \tag{1.8}$$

This expression is valid at $\alpha \ll 1$, but $v\alpha^3 \gg 1$. The last inequality is required for rapid oscillation of phase in the right side of Eq.(1.8) that allows to apply the saddle-point method. This inequality follows from the general condition $|f(\alpha)| \gg 1$ for applicability of Eq.(1.3).

1.5 THE REGION $x < v$, $v \gg 1$

If now $x < v$, the saddle point $t = \alpha$, determined from the equation $\cos \alpha = v / x$, has imaginary value. Changing $\alpha \to i\alpha$, we obtain the equation $\cosh \alpha = v / x$. Now instead of two saddle points on the real axis of the variable t we obtain two saddle points on the imaginary axis of t : $t = \pm i\alpha$. Thus one point is placed above the origin while the other is placed below the origin (and at the same distance from the origin). We must move the integration path from the real axis down to the complex plane of the variable t . Indeed, we have there

$$f''(t = -i\alpha) = -ix \sin(-i\alpha) = -\sinh \alpha < 0,$$

so that at the saddle point $t = -i\alpha$ we have a maximum of the integrand with respect to moving in the horizontal direction from this point. Then it follows from Eq. (1.1)

$$J_v(x) = \sqrt{\frac{1}{2\pi x \sinh \alpha}} \; \exp[x \sinh \alpha - v\alpha].$$

Substituting $x \sinh \alpha = v \tanh \alpha$, we obtain the second *Debye expansion* for the Bessel function

$$J_v\left(\frac{v}{\cosh \alpha}\right) = \sqrt{\frac{1}{2\pi v \tanh \alpha}} \; \exp[v (\tan \alpha - \alpha)]. \qquad (1.9)$$

1.6 THE REGION $x \ll v$

Putting $\alpha \gg 1$ in Eq. (1.9) we find the simpler expression

$$J_v\left(\frac{v}{\cosh \alpha}\right) = \sqrt{\frac{1}{2\pi v}} \; \exp[v (1 - \alpha)].$$

Here the quantity α is expressed via the argument x by means of the approximate relation

$$v = \frac{1}{2} x \exp \alpha, \quad \alpha = \ln\left(\frac{2v}{x}\right).$$

Thus, we obtain, under the conditions $x \ll v$, $v \gg 1$:

$$J_v(x) = \sqrt{\frac{1}{2\pi v}} \left(\frac{ex}{2v}\right)^v. \tag{1.10}$$

This is the first term of the Taylor expansion for the Bessel function with respect to the variable x. It is valid if the argument x of the Bessel function is small compared to its index v. It should be noted that Eq. (1.10) is also correct at large values of the argument $x \gg 1$, when $x \ll v$.

Equation (1.10) can also be derived from the basic integral expression (1.1) by means of the expansion of the exponent $\exp(ix \sin t)$ in a Taylor series with respect to x up to the vth term (all previous terms of the Taylor series give zero contribution to the integral (1.1)). Thus we make the substitution

$$\exp(ix \sin t) \rightarrow \frac{1}{v!} (ix \sin t)^v.$$

Writing further

$$i \sin t = \frac{1}{2} [\exp(it) - \exp(-it)],$$

we take into account only one nonzero term in the expression

$$\frac{x^v}{v!} (i\sin t)^v \rightarrow \frac{1}{v!} \left(\frac{x}{2}\right)^v \exp(ivt),$$

since other terms produce zero contribution to the integral (1.1). Hence, we obtain finally

$$J_\nu(x) = \frac{1}{\nu!}\left(\frac{x}{2}\right)^\nu.$$ (1.11)

Equation (1.10) can be obtained from Eq. (1.11) using Stirling's formula for the factorial:

$$\nu! = \sqrt{2\pi\nu}\left(\frac{\nu}{e}\right)^\nu.$$

However, unlike Eq.(1.10), Eq.(1.11) is also valid for moderate and small values of the index $\nu < 1$, in particular, at $\nu = 0$. Then we obtain, for example, from Eq. (1.11) that $J_0(0) = 1$ as it should be.

1.7 THE REGION $\nu - x \ll \nu$, $\nu \gg 1$

If $\alpha \ll 1$, but $\nu\alpha^3 \gg 1$ then we obtain from the Debye expansion (1.9)

$$J_\nu\left(\nu - \frac{1}{2}\nu\alpha^2\right) = \sqrt{\frac{1}{2\pi\nu\alpha}}\exp\left(-\frac{1}{3}\nu\alpha^3\right)$$ (1.12)

This is correct if the argument x is nearly equal to the index ν (but not too nearly!). It is seen that the Bessel function $J_\nu(x)$ with large index ν has an exponential asymptotic representation in some region of its argument x.

1.8 THE REGION $x \sim \nu \gg 1$

Equations (1.8) and (1.12) describe the Bessel function in the region where the argument x is nearly equal to the index ν, but not too nearly, so that the inequality $\nu\alpha^3 \gg 1$ is fulfilled, i.e.,

$$|x - v| \gg v^{1/3}.$$ (1.13)

Now we approximate Bessel function (1.1) in the region

$$|x - v| \leq v^{1/3}.$$

Of course, the results obtained below must coincide with Eq. (1.8) or Eq. (1.12) under the condition (1.13).

Rewriting the integral (1.1) in the form

$$J_v(x) = \frac{1}{\pi} \int_0^\pi \cos(x \sin t - vt)dt$$ (1.14)

due to the parity property of the integrand in Eq. (1.14) we find that only small values of the integration variable t are important in this integral under the condition $x \sim v \gg 1$. Hence, we can use the Taylor expansion

$$\sin t \approx t - \frac{1}{6}t^3$$

and then the integral (1.14) can be written in the form

$$J_v(x) = \frac{1}{\pi} \int_0^\pi \cos[(x - v)t - xt^3/6]dt.$$ (1.15)

Let us remember that the so-called *Airy function* is determined by the integral expression

$$Ai(x) \equiv \frac{1}{\pi} \int_0^\infty \cos\left(xt + \frac{1}{3}t^3\right)dt.$$ (1.16)

Comparing the integrals (1.15) and (1.16) with each other we find the approximate expression of the Bessel function via the Airy function

$$J_v(x) = \left(\frac{2}{x}\right)^{1/3} Ai\left[\frac{v-x}{(x/2)^{1/3}}\right]$$

Due to the proximity of the values of x and v to each other we finally obtain

$$J_v(x) = \left(\frac{2}{v}\right)^{1/3} Ai\left[\frac{v-x}{(v/2)^{1/3}}\right]. \qquad (1.17)$$

Equation (1.17) reduces to Eq. (1.12) under the condition $v - x \gg v^{1/3}$ and to Eq. (1.8) under the condition $x - v \gg v^{1/3}$ as it should do.

1.9 THE VALUE OF $J_v(v)$ AT $v \gg 1$

In the particular case when the argument x of the Bessel function is exactly equal to its index v, i.e. $x = v \gg 1$ we find from Eq. (1.17)

$$J_v(v) = \left(\frac{2}{v}\right)^{1/3} Ai(0). \qquad (1.18)$$

The value of Ai (0) can be derived from Eq. (1.16):

$$Ai(0) = \frac{1}{\pi}\int_0^\infty \cos\left(\frac{1}{3}t^3\right)dt = \frac{1}{2\pi}\int_0^\infty \exp\left(\frac{1}{3}it^3\right)dt + c.c. \qquad (1.19)$$

Changing the integration variable, $t = (3iu)^{1/3}$, we obtain from Eq. (1.19)

$$Ai(0) = \frac{1}{6\pi}(3i)^{1/3}\int_0^\infty \frac{\exp(-u)}{u^{2/3}}du + c.c.$$

or

$$Ai(0) = \frac{\Gamma(1/3)}{2\pi \cdot 3^{2/3}}\left[\exp\left(\frac{i\pi}{6}\right) + c.c.\right]$$

Using the relation

$$\Gamma(1/3)\Gamma(2/3) = \frac{\pi}{\sin(\pi/3)}$$

we can write the previous expression also in the form

$$Ai(0) = \frac{1}{3^{2/3}\Gamma(2/3)}.$$

Thus we finally obtain

$$J_v(v) = \frac{1}{3^{2/3}\Gamma(2/3)}\left(\frac{2}{v}\right)^{1/3} \approx \frac{0.45}{v^{1/3}}. \tag{1.20}$$

This result is applicable under the condition $v \gg 1$.

1.10 FIRST MAXIMUM VALUE OF THE BESSEL FUNCTION

The value of $J_v(v)$ derived in the previous section is not a maximum value of this Bessel function. Now we wish to calculate the maximum value by two methods. Let us consider first a simple approximate approach. We put the argument of the cosine equal to zero in Eq. (1.8), i.e. $v\alpha^3$ According to the same Eq. (1.8) we can express α via x:

$$\alpha = \left[\frac{x - v}{(v / 2)} \right]^{1/2}.$$

Thus we obtain for the position of the first maximum:

$$x - v = \frac{1}{2} \left(\frac{3\pi}{4} \right)^{2/3} v^{1/3} \approx 0.88 v^{1/3}. \tag{1.21}$$

It is seen that the maximum value of the Bessel function is achieved when the argument x exceeds its index v. According to Eq. (1.8) the maximum value of the Bessel function is given by the expression

$$J_v(x)\big|_{max} = \sqrt{\frac{2}{\pi v \alpha}} = \sqrt{\frac{2}{\pi v (3\pi / 4v)^{1/3}}} \approx \frac{0.69}{v^{1/3}}. \tag{1.22}$$

The exact position for the first maximum of the Bessel function when $v \gg 1$ can be determined using Eq. (1.17). Obviously, the derivative of the Airy function vanishes in this point, i.e.

$$\text{Ai'} \left[\frac{v - x}{(v / x)^{1/3}} \right] = 0. \tag{1.23}$$

It follows from the numerical tables of the Airy function that the derivative of the Airy function vanishes for the first time at the value of the argument ≈ -1.0.

Thus it follows from Eq. (1.23) that

$$x - v \approx \left(\frac{1}{2} v \right)^{1/3} \approx 0.79 v^{1/3}. \tag{1.24}$$

It is seen that the exact result (1.24) is in good agreement with the approximate expression (1.21).

A more exact (compared to Eq. (1.22)) maximum value of Bessel function is derived from Eq.(1.17) using the numerical value of the Airy function $Ai(-1.0) \approx -0.535$.

Thus we find

$$J_v(x)\Big|_{max} \approx \left(\frac{2}{v}\right)^{1/3} Ai(-1.0) \approx \frac{0.67}{v^{1/3}}. \tag{1.25}$$

It is again seen that the exact result (1.25) and the approximate result (1.22) are in good agreement with one other.

In Fig.1 we show the Bessel function $J_8(x)$ as an example. The regions are marked where this function is described by the elementary expressions obtained above. These expressions are in agreement with one other on the boundaries of their applicability.

This is a typical approach for the analysis of any higher transcendental function represented in integral form.

PROBLEMS

Problem 1. Using the Taylor expansion

$$F(x) \approx F(n+1/2) + F'(n+1/2)x + F''(n+1/2)x^2/2$$

obtain the so called *Euler – MacLaurin* formula

$$\int_0^\infty F(x)dx \approx \sum_{n=0}^\infty F(n+1/2) - \frac{1}{24}F'(0).$$

Problem 2. Obtain an approximate expression for the *full elliptic integral of the first kind* as $k \to 1$:

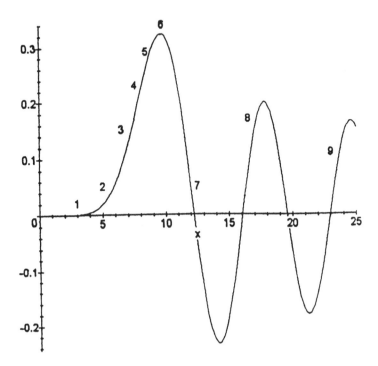

Figure 1. Approximation of the Bessel function $J_8(x)$ in various regions of the variable x. Numbers on the curves correspond to the numbers of equations in the text: 1 ~ (1.10); 2 ~ (1.9); 3 ~ (1.12); 4 ~ (1.20); 5 ~ (1.17); 6 ~ (1.25); 7 ~ (1.8); 8 ~ (1.6) and 9 ~ (1.7).

$$K(k) = \int_0^{\pi/2} \frac{d\varphi}{\sqrt{1 - k^2 \sin^2 \varphi}} \approx \ln \frac{4}{\sqrt{1 - k^2}}.$$

Hint:. In the process of this derivation divide the integration interval into two parts: $[0, \varphi_0]$ and $[\varphi_0, \pi/2]$, where the angle φ_0 is near to $\pi/2$. In the first integration interval you can put $k = 1$, while in the second interval the following substitution can be made

$$1 - k^2 \sin^2 \varphi \approx (1 - k^2) + \phi^2 , \quad \phi \equiv \pi / 2 - \varphi .$$

The arbitrary angle φ_0 disappears in the final result. This approach is typical of calculations of integrals giving logarithmic functions.

$$J_v (x, y) \equiv \sum_{k=-\infty}^{\infty} J_{v-2k}(x)J_k(y)$$

Problem 3. The so-called *generalized Bessel functions* are used in various physical problems. Their simple representations under the condition $v \gg 1$ are still unknown today. Obtain possible representations using the methods described in this chapter for the usual Bessel functions and consider various parts of the (x,y) plane.

Problem 4. Using the next order of the saddle-point method, calculate the correction to Stirling's formula when $x \gg 1$:

$$\Gamma(x + 1) = \int_x^{\infty} t^x \exp(-t)dt \approx \sqrt{2\pi x} \left(\frac{x}{e} \right)^x \left(1 + \frac{1}{12x} \right)$$

Explain the high accuracy of this formula when $x \sim 1$ (for example, the right side of this expression for $1! = \Gamma(2) = 1$ is equal to 0.995 and for $2! = \Gamma(3) = 2$ it is equal to 1.9994).

Problem 5. Using the integral representation for Legendre polynomials

$$P_n(\cos \theta) = \frac{\sqrt{2}}{\pi} \int_0^{\theta} \frac{\sin[(n + 1/2)\varphi]}{\sqrt{\cos \varphi - \cos \theta}} d\varphi \quad (0 < \theta < \pi)$$

obtain an asymptotic representation for these polynomials when

$$n \gg 1, \quad n \gg \frac{1}{\theta}, \quad n \gg \frac{1}{\pi - \theta} :$$

$$P_n(\cos\theta) \approx \sqrt{\frac{2}{\pi(n+1/2)\sin\theta}}\, \sin\left[(n+1/2)\theta + \frac{\pi}{4}\right].$$

Problem 6. Using the integral representation from the previous problem, obtain the asymptotic representation for Legendre polynomials when $n \gg 1$ via the Bessel function:

$$P_n(\cos\theta) \approx J_0\left[\left(n+\frac{1}{2}\right)\theta\right].$$

Problem 7. Using the integral representation for *Hermite polynomials*

$$H_n(x) = \sqrt{\frac{2^n}{\pi}} \int_{-\infty}^{\infty} (x + it)^n \exp(-t^2)dt,$$

obtain the asymptotic representation for these polynomials at $n \gg 1$:

$$H_n(x) \approx (-2)^n (2n-1)!! \cos\left(x\sqrt{4n+1}\right)\exp(x^2/2);$$

$$H_{2n+1}(x) \approx (-2)^n \sqrt{2}(2n-1)!! \sin\left(x\sqrt{4n+3}\right)\exp(x^2/2).$$

Problem 8. Using the integral representation for the *McDonald function* (modified Bessel function)

$$K_\nu(x) = \int_0^\infty \exp(-x\cosh t)\cosh(\nu t)dt,$$

obtain two asymptotic representations for this function:

(1) $x \gg 1: \quad K_v(x) \approx \sqrt{\dfrac{\pi}{2x}} \exp(-x);$

$v \gg 1: \quad K_v(vx) \approx$

(2)

$$\approx \frac{\sqrt{\pi / 2v}\left(1 + \sqrt{1 + x^2}\right)^v}{x^v\left(1 + x^2\right)^{1/4}} \exp\left\{-v\left[\sqrt{1 + x^2}\right]\right\}.$$

Chapter 2

Continual Integrals

2.1 TEMPORAL GREEN'S FUNCTIONS

In this chapter we consider the exact and approximate derivation of so-called *continual* (i.e. infinite multiple) integrals for the example of the temporal Green's function G in quantum mechanics. This function connects the wave function $\Psi(x,t)$ of the system for time t with the wave function $\Psi(x,0)$ for initial time $t=0$:

$$\Psi(x,t) = \int G(x,x';t)\Psi(x',0)dx'. \tag{2.1}$$

According to this definition the Green's function is a dimensional quantity. However, we can introduce a normalized factor A with a dimension of length (its derivation is given below) and rewrite Eq.(2.1) in the form

$$\Psi(x,t) = \int AG(x,x';t)\Psi(x',0)\frac{dx'}{A} \tag{2.2}$$

so that the quantity AG is now dimensionless one.

According to Feynman's description of quantum mechanics the transition amplitude AG (see Eq.(2.2)) from some initial state i to the final state f can be expressed via the classical action S which connects the initial and final states of a quantum-mechanical particle:

$$AG = \sum_{i \to f} \exp\left(\frac{i}{\hbar} S\right) \tag{2.3}$$

Here $\hbar$ is the Planck constant, and the summation is taken over all classical paths connecting the fixed initial (i) and final (f) positions of the considered particle. Equation (2.3) is the postulate of quantum mechanics which is equivalent to the well-known postulate of the temporal Schroedinger equation.

The classical action S can be expressed via the Lagrange function L:

$$S = \int_0^t L(x, v)dt'. \tag{2.4}$$

Here 0 and t are the initial and final times, respectively; x is the coordinate of the particle at a given time t', and v is its velocity. For the sake of simplicity we consider the one-dimensional problem only, and we assume that the Lagrange function does not depend on time explicitly.

Dividing the whole temporal interval $[0, t]$ into a large number $N \gg 1$ of small temporal intervals with a width dt of each interval so that $t = Ndt$, we obtain from Eq.(2.4):

$$S = dt \cdot \sum_{k=1}^{N} L(x_k, x_{k-1}). \tag{2.5}$$

The velocities can be expressed via the coordinates using the obvious relations:

$$v_k = \frac{x_k - x_{k-1}}{dt}.$$

Substituting Eq.(2.5) into Eq.(2.3) we obtain the continual (i.e. infinite multiple) integral determining the quantum-mechanical transition amplitude:

$$G(x,v,t) = \iint \ldots \int\limits_{-\infty}^{\infty} \exp\left\{\frac{idt}{\hbar} \sum_{k=1}^{N} L(x_k, x_{k-1})\right\} \frac{dx_1}{A} \frac{dx_2}{A} \ldots \frac{dx_{N-1}}{A} \frac{1}{A}.$$

(2.6)

Here $x_N = x$.

The intermediate coordinates $x_1, x_2, \ldots, x_{N-1}$ of a particle take arbitrary values in the interval $[-\infty, \infty]$ (see Fig. 2). The quantity A is the normalized factor for each of the coordinates x_k. The derivation of this factor is given later. One of the typical possible paths is depicted in Fig. 2. For the sake of simplicity we put $x_0 = x' = 0$ for the initial coordinate of the particle.

The calculation of each of the $(N - 1)$ integrals in Eq.(2.6) determines the solution of our problem: the derivation of the transition amplitude $G(x, t)$ from the initial point $x' = t = 0$ to the final point x,t on the coordinate-time plane (see Fig. 2). Of course this calculation depends on the concrete form of the Lagrange function. Below we consider two simple examples of this derivation: (1) free particle, and (2) harmonic oscillator.

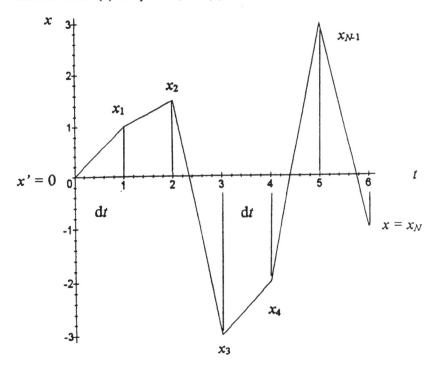

Figure 2. One of the possible paths connecting fixed initial $(x' = t = 0)$ and final (x, t) points.

2.2 A FREE QUANTUM MECHANICAL PARTICLE

Let us calculate first of all the quantum-mechanical transition amplitude for a free particle from the origin where this particle is placed at the moment time $t = 0$ to the some point (x,t). According to the rules of quantum mechanics the square of the modulus of this amplitude determines the probability of the transition.

The Lagrange function of a free particle coincides with its kinetic energy, i.e.

$$L(v) = \frac{m}{2}v^2 = \frac{m(x_k - x_{k-1})^2}{2(dt)^2}. \tag{2.7}$$

We choose the system of units so that the mass of the particle is $m = 1$ and the Planck constant is $\hbar = 1$. Substituting Eq.(2.7) into Eq.(2.6) we begin the consequent derivation of the $(N - 1)$ integrals.

The coordinate x_1 is contained in the next two terms of the sum over k :

$$\frac{1}{2}x_1^2 + \frac{1}{2}(x_2 - x_1)^2 = (x_1 - x_2/2)^2 + \frac{1}{4}x_2^2. \tag{2.8}$$

The derivation of Poisson's integral over dx_2 produces the first factor in the transition amplitude (2.6):

$$a_1 = \frac{\sqrt{i\pi dt}}{A}. \tag{2.9}$$

The coordinate x_2 is contained in the next two terms of the sum over k in the exponent of Eq.(2.6) (taking into account the expression (2.8)):

$$\frac{1}{4}x_2^2 + \frac{1}{2}(x_3 - x_2)^2 = \frac{3}{4}(x_2 - 2x_3/3)^2 + \frac{1}{6}x_3^2. \tag{2.10}$$

The derivation of Poisson's integral over dx_2 produces the second factor in the transition amplitude (2.6):

$$a_2 = \frac{1}{A}\sqrt{\frac{4i\pi dt}{3}}.$$

The coordinate x_3 is contained in the next two terms of sum over k in the exponent of Eq.(2.6) (taking into account the expression (2.10)):

$$\frac{1}{6}x_3^2 + \frac{1}{2}(x_4 - x_3)^2 = \frac{4}{6}(x_3 - 3x_4/4)^2 + \frac{1}{8}x_4^2. \qquad (2.11)$$

The derivation of Poisson's integral over dx_3 produces the third factor in the transition amplitude (2.6):

$$a_3 = \frac{1}{A}\sqrt{\frac{6i\pi dt}{4}}. \qquad (2.12)$$

Generalising Eq. (2.11) to arbitrary values of k we obtain that the coordinate x_k is contained in the combination:

$$\frac{k+1}{2k}\left(x_k - \frac{k}{k+1}x_{k+1}\right)^2. \qquad (2.13)$$

It is seen from Eq.(2.13) that Poisson's integral over dx_k produces the next factor in the transition amplitude (2.6):

$$a_k = \frac{1}{A}\sqrt{\frac{2ki\pi dt}{k+1}}. \qquad (2.14)$$

Thus, multiplying all factors of the type (2.14) we derive the transition amplitude (2.6):

$$G(x,t) = a_1 a_2 ... a_{N-1} \frac{1}{A}. \qquad (2.15)$$

In the process of multiplying the factors $k/(k+1)$ in Eq.(2.15) the numerators and denominators cancel from neighbouring terms. For example,

$$a_{k-1} a_k a_{k+1} = \sqrt{\frac{k-1}{k+2} \frac{(2\pi i dt)^{3/2}}{A^3}}. \tag{2.16}$$

Hence, we find from Eq.(2.15)

$$G(x,t) = \frac{(2\pi i dt)^{(N-1)/2}}{A^N \sqrt{N}} \exp\left(\frac{ix^2}{2Ndt}\right) \tag{2.17}$$

Writing $Ndt = t$ and reconstructing the variable x' we find from Eq.(2.17)

$$G(x,x';t) = \frac{(2\pi i dt)^{(N-1)/2}}{A^N \sqrt{N}} \exp\left[\frac{i(x-x')^2}{2t}\right]. \tag{2.18}$$

The normalized factor A in this expression can be derived by letting the time $t \to 0$. Then according to Eq.(2.1) and the condition of continuity of the wave function the integration of Eq.(2.18) over dx' should produce unity. Indeed, the wave function in the right side of Eq.(2.1) is a smooth function so that we can cancel it with the wave function in the left side of Eq.(2.1). Thus, calculating the simple Poisson's integral, we find from Eq.(2.18) the relation

$$1 = \frac{(2\pi i dt)^{(N-1)/2}}{A^N \sqrt{N}} \sqrt{2\pi i dt}. \tag{2.19}$$

From Eq.(2.19) we derive the normalized factor A:

$$A = \sqrt{2\pi i dt}. \tag{2.20}$$

Substituting Eq.(2.20) into Eq.(2.18) we obtain the temporal Green's function

$$G(x, x', t) = \frac{1}{\sqrt{2\pi i t}} \exp\left[\frac{i(x - x')^2}{2t}\right].$$
(2.21)

The derivation of temporal Green's functions using continual integration for particles in some potential field is performed analogously. However, the derivation is more cumbersome compared to the case of a free particle. The method of calculation of the normalized factor A is the same as above, but its value, of course, does not coincide with Eq.(2.20). Further we consider as an example the case of a particle in the field of a harmonic oscillator potential.

2.3 HARMONIC OSCILLATOR

Let us calculate the transition amplitude for a particle in the field of a harmonic oscillator potential. The Lagrange function is now of the form

$$L(x, v) = \frac{1}{2}v^2 - \frac{1}{2}\omega^2 x^2.$$
(2.22)

We again put the mass of a particle to be equal to unity for the sake of simplicity.

The classical path corresponds to vibrational motion so that the coordinate and the velocity of a particle are of the well-known form

$$x_c(t) = C \sin \omega t, \quad v_c(t) = C\omega \cos \omega t.$$
(2.23)

For the sake of simplicity we have chosen paths which begin from the origin $x' = t' = 0$.

According to Eq.(2.3) the temporal Green's function is given by the relation:

$$G(x, t) \propto \sum_{i \to f} \exp[iS(x, v)],$$
(2.24)

where the Planck constant is put equal to unity as in the previous section. The summation in Eq.(2.24) is taken over all paths connecting the fixed initial point $x' = t' = 0$ and final point x, t. The classical action S is presented as an integral involving the Lagrange function

$$S(x, v) = \int\limits_0^t L[x(\tau), v(\tau)]d\tau. \qquad (2.25)$$

Substituting Eq.(2.23) into Eq.(2.22) we derive the Lagrange function for an arbitrary classical path:

$$L(x_C, v_C) = \frac{1}{2} C^2 \omega^2 \cos(2\omega t). \qquad (2.26)$$

Substituting Eq.(2.26) into Eq.(2.25) we derive the action for the classical path

$$S_c = \frac{1}{4} C^2 \omega \sin(2\omega t) = \frac{\omega \cos \omega t}{2 \sin \omega t} x_c^2. \qquad (2.27)$$

We have expressed here the action via the time and coordinate for a classical path (see Eq. (2.23)).

Let us now present (following to R. Feynman) the coordinate of a particle in the next form $x = x_c + y$, where the quantity y determines the deviation of a given path from the classical path (this quantity is not small compared to x_c). Substituting this relation into Eq.(2.25) we obtain

$$S = S_c + \int\limits_0^t \left(\frac{1}{2} v_y^2 - \frac{1}{2} \omega^2 y^2 \right) d\tau. \qquad (2.28)$$

The terms of first order in y produce zero contribution to the action (see below). Therefore they are absent in Eq.(2.28).

Substituting Eqs.(2.27) and (2.28) into Eq.(2.24) we obtain the temporal Green's function in the form

$$G(x,t) = T(t)\exp\left\{\frac{ix^2\omega\cot\omega t}{2}\right\},$$

(2.29)

$$T(t) \propto \sum_{i \to f} \exp\left\{i\int_0^t\left(\frac{1}{2}\overset{.}{v}_y^2 - \frac{1}{2}\omega^2 y^2\right)d\tau\right\}.$$

In the second formula of Eq.(2.29) the summation is made over all paths which begin and finish on the classical path according to the definition of y. Thus, we change trajectories. While the above all paths begin at the point $x' = t' = 0$, and finish at the point (x, t), we have now $y(0) = y(t) = 0$ for all paths.

Hence the function $y(t)$ is periodic with period t. Therefore we can expand this function in a Fourier series

$$y(\tau) = \sum_{n=1}^{N \to \infty} y_n \sin\left(\frac{\pi n}{t}\tau\right)$$

(2.30)

Due to the above boundary conditions this expansion contains only sine functions.

Instead of the variation of the function $y(\tau)$ we shall vary the coefficients of the Fourier expansion (2.30). Of course, the normalized factor is different from the normalized factor for y.

Substituting Eq.(2.30) into the second formula of Eq.(2.29), we find

$$\int_0^t y^2(\tau)d\tau = \sum_{n,m} y_n y_m \int_0^t \sin\left(\frac{\pi n}{t}\tau\right)\sin\left(\frac{\pi m}{t}\tau\right)d\tau = \frac{t}{2}\sum_n y_n^2. \quad (2.31)$$

Analogously we obtain

$$\int_0^t \overset{.}{y}^2(\tau)d\tau = \frac{\pi^2}{2t}\sum_n n^2 y_n^2. \quad (2.32)$$

Substituting Eqs. (2.31) and (2.32) into the second formula of Eq.(2.29) we obtain

$$T(t) = \iint \ldots \int\limits_{-\infty}^{\infty} \exp\left\{\frac{it}{4}\sum_{n=1}^{N\to\infty}\left(\frac{\pi^2 n^2}{t^2} - \omega^2\right)y_n^2\right\}\frac{dy_1}{B}\frac{dy_2}{B}\ldots\frac{dy_N}{B}.$$

(2.33)

Here B is the normalized factor for the Fourier coefficients y_n, which is analogous to the normalised factor A in the previous Section.

Let us calculate the infinite multiple integral in Eq.(2.33). It is equal to the product of N single Poisson integrals. The typical nth integral is of the form

$$I_n = \frac{1}{B}\sqrt{\frac{4\pi i}{\left(\dfrac{\pi^2 n^2}{t^2} - \omega^2\right)}}.$$

(2.34)

Hence,

$$T(t) = I_1 I_2 \ldots I_N, \qquad N \to \infty.$$

(2.35)

In order to derive this infinite product we use the known mathematical formula

$$\prod_{n=1}^{\infty}\left(1 - \frac{x^2}{\pi^2 n^2}\right) = \frac{\sin x}{x}.$$

(2.36)

This can be obtained, for example, from representations of the gamma function. Taking into account Eq.(2.36), we substitute Eq.(2.34) into Eq.(2.35)

$$T(t) = \sqrt{\frac{\omega t}{\sin \omega t}}\left(\frac{4it}{\pi B^2}\right)^{N/2}\frac{1}{N!}.$$

(2.37)

The normalized factor B in this dependence can be derived by means of the limit of a free particle, i.e. $\omega \to 0$. Comparing Eqs. (2.37) and (2.21) we obtain the following equation for B:

$$2\pi i t = \left(\frac{\pi B^2}{4it} \right)^N (N!)^2.$$ (2.38)

Hence, from Eqs.(2.37) and (2.38) we have the final simple result

$$T(t) = \sqrt{\frac{\omega}{2\pi i \sin \omega t}}.$$ (2.39)

Thus the normalized Green's function of the harmonic oscillator is of the form (we use Eq.(2.39) and the first formula of Eq.(2.29)):

$$G(x,t) = \sqrt{\frac{\omega}{2\pi i \sin \omega t}} \, \exp\left\{ \frac{i\omega x^2 \cot \omega t}{2} \right\}.$$ (2.40)

The expression (2.40) reduces to Eq.(2.21) in the limiting case of a free particle as it should do.

Of course if the initial coordinate x' is nonzero the Green's function is of a more difficult form since it depends only on $(x - x')$ for a free particle. The reason is that the external field violates the invariance with respect to the shift of the reference frame.

It should be noted that it follows from Eqs.(2.31) and (2.32) that the integration of linear terms on the variable y produces a zero contribution.

2.4. THE SADDLE-POINT METHOD

If the classical action is large compared to the Planck constant then the so-called *WKB approximation* is appropriate. In this case the continual integral (2.3) can be calculated in the general form for a particle in a potential field by the saddle-point method (see Chapter 1). Let us rewrite the Lagrange function via discrete values of variables:

$$L(x_k - x_{k-1}) = \frac{(x_k - x_{k-1})^2}{2(dt)^2} - U(x_k). \tag{2.41}$$

Here the mass of the particle is equal to unity as above. The quantity U is the potential energy of the particle.

According to the saddle-point method we find the position of the saddle point in the integration over the variable x_k. This variable is contained in two terms of the sum over k in Eq.(2.5). Therefore we obtain

$$\frac{\partial L(x_k - x_{k-1})}{\partial x_k} + \frac{\partial L(x_{k+1}, x_k)}{\partial x_k} = 0. \tag{2.42}$$

Substituting Eq.(2.41) into Eq.(2.42) we obtain

$$-\frac{\partial U}{\partial x_k} = \frac{x_{k+1} + x_{k-1} - 2x_k}{(dt)^2} = a_k. \tag{2.43}$$

Here a_k is the acceleration of the particle in the kth temporal interval. The relation (2.43) is obviously Newton's second law which determines the classical path of the considered particle. We denote the coordinate for this path as $x_{kc} = x_c(t_k)$, $t_k = kdt$.

Thus we obtain the WKB-expression for Green function (with pre-exponential accuracy):

$$G(x,t) \propto \exp\left\{ idt \sum_{k=1}^{N} L(x_{kc}, x_{(k-1)c}) \right\}. \tag{2.44}$$

Finally replacing the sum by an integral we find from Eq.(2.44):

$$G(x,t) \propto \exp\left\{ \frac{i}{\hbar} \int_0^t L[x_c(t'), v_c(t')]dt' \right\}. \tag{2.45}$$

For example, in the case of a free particle we have $L=E$, where the energy E is a constant. Hence it follows from Eq.(2.45) that

$$G(x,t) \propto \exp(iEt). \qquad (2.46)$$

Obviously for a free particle we have $E = v^2/2$, where the velocity is $v = x/t$, or in the more general form $v = (x - x')/t$ if the initial coordinate x' is nonzero. Hence, the Lagrange function for a classical path can be written in the form

$$L_c = \frac{(x - x')^2}{2t}$$

and Eq.(2.21) follows from Eq.(2.46) with pre-exponential accuracy as it should do.

Analogously we can obtain the exponent of Eq.(2.27) for the case of a harmonic oscillator using the WKB expression (2.46).

The Green's function determined by Eq.(2.1) can also be written via the eigenfunctions $\varphi_n(x)$ of the Schroedinger equation with the Hamiltonian $H=T+U$, where T is the kinetic energy operator, and U is the potential energy of the considered particle. These eigenfunctions satisfy the differential equation

$$H\varphi_n(x) = E_n\varphi_n(x). \qquad (2.47)$$

Here E_n are the quantum-mechanical eigenvalues of the energy. It follows from Eqs.(2.1) and (2.47) that the Green's function can be expressed via the eigenfunctions of the problem:

$$G(x,x';t) = \sum_n \varphi_n(x)\varphi_n^*(x')\exp(-iE_nt). \qquad (2.48)$$

Now we can obtain the WKB expression (2.45) by the second method. We must substitute into Eq.(2.48) the WKB expressions for the eigenfunctions which are of the form

$$\varphi_n(x) \propto \exp\left[i\int\limits_a^x p_n(x'')dx'' \right], \quad \varphi_n(x') \propto \exp\left[i\int\limits_a^{x'} p_n(x'')dx'' \right].$$

(2.49)

Here a is the classical turning point, and p_n is the classical momentum of the particle, i.e.

$$p_n(x) = \sqrt{2[E_n - U(x)]}.$$

It follows from Eqs.(2.49) and (2.48) that the WKB expression for the Green's function is

$$G(x,x';t) \propto \sum_n \exp\left\{ i\int\limits_{x'}^x p_n(x'')dx'' - iE_n t \right\}.$$

(2.50)

This expression is equivalent to Eq.(2.45) (if we put $x' = 0$). The exponent of Eq.(2.50) contains the classical action for a real classical path of the particle between fixed initial and final points.

Unlike Eq.(2.45) the second method allows us to obtain also the correct pre-exponential factor for the Green's function, Eq.(2.50). In order to derive this factor we should use the known pre-exponential factors in the WKB eigenfunctions, Eq. (2.49).

PROBLEMS

Problem 1. Calculate the temporal Green function for a harmonic oscillator perturbed by an external force $f(t)$ using the method of continual integration.
Answer ($m = \hbar = 1$):

$$G(x,t) = \sqrt{\frac{\omega}{2\pi i}} \exp(iS_c);$$

$$S_c = \frac{1}{2}\omega x^2 \cot \omega t + \frac{x}{\sin \omega t}\int_0^t f(t')\sin \omega t'dt'$$

$$-\frac{1}{\omega \sin \omega t}\int_0^t dt'f(t')\sin[\omega(t-t')]\int_0^{t'} dt'f(t'')\sin(\omega t'').$$

Problem 2.   Calculate the temporal Green's function for a harmonic oscillator by the method of continual integration in the case $x' \neq 0$.
Answer:

$$G(x,x;t) = \sqrt{\frac{\omega}{2\pi i \sin \omega t}}\,\exp(iS_c);$$

$$S_c = \frac{1}{2}\omega(x^2 + x'^2)\cot \omega t - \frac{\omega x \cdot x'}{\sin \omega t}.$$

Check that this expression reduces Eq.(2.40) in the case $x' = 0$. Check also that exact and WKB results coincide.

Problem 3. Derive the WKB penetration coefficient of the one-dimensional potential barrier

$$D = \exp\left(-\frac{2}{\hbar}\int_a^b \sqrt{2m(U(x) - E)}dx\right)$$

by the method of continual integration assuming that this coefficient is exponentially small. Use the saddle-point method. Here $U(a) = U(b) = E$.

Problem 4. Calculate the continual integral

$$I = \lim_{N \to \infty}\int_0^1\int_0^1...\int_0^1 f\left(\frac{x_1 + x_2 + ... + x_N}{N}\right)dx_1 dx_2 ... dx_N,$$

where $f(0) = f(1)$. Use a Fourier expansion for the integrand.
Answer:

$$I = f(1/2).$$

Chapter 3

Calculation of Green's Functions

3.1 GREEN'S FUNCTIONS

The temporal Green's function was introduced in the previous chapter. This function is useful in the solution of various physical problems. Instead of a general definition of the Green's function we restrict ourselves to some examples from quantum mechanics (as in the previous chapter). Thus we investigate here solutions of the Schroedinger equation:

$$(H - E)\psi(\mathbf{r}) = 0.$$

Here H is the differential operator (Hamiltonian) with respect to the coordinate $\mathbf{r}$, and E is the energy. The quantity $\psi(\mathbf{r})$ is the wave function. The Green's function can be determined as a solution of the equation

$$\left(H(\mathbf{r}) - E \right) G_E(\mathbf{r}, \mathbf{r}') = \delta(\mathbf{r} - \mathbf{r}'). \tag{3.1}$$

The general expression for the Green's function can be written in the simple form

$$G_E(\mathbf{r}, \mathbf{r}') = \sum_i \frac{\varphi_i^*(\mathbf{r})\varphi_i(\mathbf{r}')}{E_i - E}. \tag{3.2}$$

Here the eigenfunctions of the differential operator φ_i are satisfied by the equation

$$(H - E_i)\varphi_i(\mathbf{r}) = 0. \tag{3.3}$$

E_i is the energy eigenvalue which corresponds to the eigenfunction $\varphi_i(\mathbf{r})$.

Indeed, if we substitute Eq.(3.2) into Eq.(3.1) and use Eq.(3.3) then we obtain the well-known condition of completeness for the eigenfunctions:

$$\sum_i \varphi_i(\mathbf{r})\varphi_i(\mathbf{r'}) = \delta(\mathbf{r} - \mathbf{r'}). \tag{3.4}$$

It is seen that the Green's function, Eq.(3.2), is the Fourier component of the temporal Green's function, Eq.(2.48) from the previous chapter.

It follows from Eq.(3.2) that the Green's function is an analytical function of the complex energy variable E. In addition, this function has simple poles at the discrete eigenvalues of energies $E = E_i$. The continuous values of energy produce a cut along the real axis of the complex energy E.

The definition (3.2) is inconvenient in practical calculations due to the infinite sum over eigenvalues. We consider the most interesting cases when the Green function can be derived in closed form.

3.2 STANDARD METHOD OF DERIVATION OF GREEN'S FUNCTIONS

We consider now an arbitrary linear differential equation of the second order depending only on one radial variable:

$$H = \frac{d^2}{dr^2} + q(r). \tag{3.5}$$

Let us consider two solutions of the equation

$$Hu(r) = Eu(r), \tag{3.6}$$

which satisfy, on the boundaries of the range of independent variable $[r_1, r_2]$, boundary conditions of the general form

$$a_1 u_1(r_1) + a_2 u_1'(r_1) = 0;$$

$$b_1 u_2(r_2) + b_2 u_2'(r_2) = 0. \tag{3.7}$$

These solutions are not eigenfunctions of the operator (3.5) since for each of these functions only one boundary condition is satisfied instead of two. Therefore the quantity E has an arbitrary value, and this is not an energy eigenvalue. The corresponding equation for the Green's function is of the form

$$(H - E)G_E(r,r') = \delta(r - r').$$ (3.8)

Its solution can be written in the simple form:

$$G_E(r,r') = u_1(r_<)u_2(r_>).$$ (3.9)

The quantities $r_<$, $r_>$ are the smaller and larger values of the variables r, r' respectively. This notation will often be used below.

The function $u_1(r_<)$ satisfies the first of the boundary conditions (3.7) while the function $u_2(r_>)$ satisfies the second condition of Eq.(3.7). Indeed, if $r \neq r'$, then substitution of Eq.(3.9) into Eq.(3.8) produces Eq.(3.6) for the eigenfunctions. Further, integrating Eq.(3.8) over the variable r in the small vicinity of the quantity r' we obtain

$$\int_{r'-\varepsilon}^{r'+\varepsilon} \frac{d^2}{dr^2}\{u_1(r_<)u_2(r_>)\}dr = 1.$$ (3.10)

Here $\varepsilon \rightarrow +0$. After integration we obtain

$$u_1(r)u_2'(r) - u_1'(r)u_2(r) = 1,$$ (3.11)

where we have changed $r' \rightarrow r$ in Eq.(3.11) after integration.

The left side of Eq.(3.11) is the so-called *Wronskian* of the solutions of the differential equation (3.6). This Wronskian is equal to a constant (unity in our case). This follows from Eq.(3.6).

Let us emphasis once more that these functions u_1, u_2 are not eigenfunctions of the Hamiltonian, but are two linearly independent solutions of the differential equation (3.6) for the same value of the energy E, which corresponds to different boundary conditions.

3.3 COULOMB GREEN'S FUNCTIONS

In this section we derive the *Coulomb Green's function*, i.e. the Green's function of the three-dimensional Schroedinger equation with a Coulomb potential. The Hamiltonian of the hydrogen atom is of the form

$$H = -\frac{1}{2}\Delta - \frac{1}{r}. \tag{3.12}$$

Here we use the atomic system of units so that the electron mass, the electron charge and the Planck constant are equal to unit.

The potential depends only on the radial variable. Therefore we can expand the Green's function over partial waves:

$$G_E(\mathbf{r}, \mathbf{r}') = \sum_{l,m} G_{El}(r, r') \frac{1}{r \cdot r'} Y_{lm}^*(\Omega) Y_{lm}(\Omega'). \tag{3.13}$$

Here $Y_{lm}(\Omega)$ is the spherical function depending on the solid angle Ω, the summation is taken over all orbital quantum numbers l and magnetic quantum numbers m. $G_{El}(r, r')$ is the radial Green's function.

Let us use the analogous expansion of the delta-function

$$\delta(\mathbf{r} - \mathbf{r}') = \frac{1}{r \cdot r'} \delta(r - r') \sum_{lm} Y_{lm}^*(\Omega) Y_{lm}(\Omega'). \tag{3.14}$$

Substituting Eqs.(3.13) and (3.14) into Eq.(3.1) we obtain the one-dimensional differential equation for the radial Green's function:

$$\left[-\frac{1}{2}\frac{d^2}{dr^2} + \frac{l(l+1)}{2r^2} - \frac{1}{r} - E \right] G_{El}(r, r') = \delta(r - r'). \tag{3.15}$$

The analogous equation for the functions $u(r)$ differs from Eq.(3.15) only by having zero on the right side instead of a delta-function.

After changing the independent variable

$$r \equiv \frac{x}{2\sqrt{-2E}}$$

we obtain the so-called *Whittaker equation*:

$$\frac{d^2 u(x)}{dx^2} + \left[\frac{v}{x} + \frac{1/4 - \mu^2}{x^2} - \frac{1}{4} \right] u(x) = 0. \qquad (3.16)$$

Here the following notation is introduced:

$$v \equiv \frac{1}{\sqrt{-2E}}, \quad \mu \equiv l + \frac{1}{2}.$$

The Whittaker equation should be solved on the interval $0 < x < \infty$. The boundary condition at $x = 0$ is that the value of $u_1(0)$ is finite. The second boundary condition is $u_2(\infty) = 0$. The so-called *Whittaker functions* satisfy the conditions:

$$u_1(x) = \text{const} \cdot M_{v,\mu}(x);$$
$$\qquad\qquad\qquad (3.17)$$
$$u_2(x) = \text{const} \cdot W_{v,\mu}(x).$$

These functions can be expressed via the so-called *degenerate hypergeometric functions.*
Thus, according to the general solution (3.9) we obtain

$$G_{El}(r, r') = C \cdot M_{v,l+1/2}\left(\frac{2r_<}{v} \right) W_{v,l+1/2}\left(\frac{2r_>}{v} \right) \qquad (3.18)$$

The constant C in this expression is derived from the condition (3.11). Since the Wronskian does not depend on the independent variable x, we can derive it as $x \rightarrow 0$. Then we find

$$M_{v,\mu}(x) \approx x^{1/2+\mu} \; :$$

$$W_{v,\mu}(x) \approx \frac{\Gamma(2\mu)}{\Gamma(1/2+\mu-v)} x^{1/2-\mu} \; .$$

Hence, the Wronskian of these functions is equal to:

$$M_{v,\mu}(x)W'_{v,\mu}(x) - M'_{v,\mu}(x)W_{v,\mu}(x) = -\frac{\Gamma(2\mu+1)}{\Gamma(1/2+\mu-v)} \; .$$

Since the differentiations with respect to the variable x and the variable r differ from one other by $(2/v)$ times, and the Wronskian of the system (3.15) is equal to $-1/2$, then the constant C in Eq.(3.18) is given by expression

$$C = v \frac{\Gamma(1/2+\mu-v)}{\Gamma(2\mu+1)} \; .$$

Substituting this value of the constant in Eq.(3.18) we finally obtain the radial Coulomb Green's function

$$G_{El}(r,r') = v \frac{\Gamma(l+1-v)}{\Gamma(2l+2)} M_{v,l+1/2}\left(\frac{2r_<}{v}\right) W_{v,l+1/2}\left(\frac{2r_>}{v}\right)$$

(3.19)

Thus we have obtained a closed expression for three three-dimensional Coulomb Green's function. Analogously we can calculate the Green's functions in other cases. The analytical results are obtained in those cases when the corresponding homogeneous differential equation is solved analytically.

3.4 STURMIAN FUNCTIONS

In this section we consider another set of eigenfunctions: the so-called *Sturmian functions*. Instead of the general approach we investigate again the example of a hydrogen-like atom with a charge Z on the nucleus. As above we use here the atomic system of units. According to the results of the previous section the eigenfunctions satisfy the equation

$$\left[-\frac{d^2}{2dr^2} + \frac{l(l+1)}{2r^2} - \frac{Z}{r} \right] u_{nl}(r) = E_n u_{nl}(r).$$

Here the eigenvalues of the energy are given by the well-known formula

$$E_n = -\frac{Z^2}{2n^2}.$$

Let us remember that the radial Schroedinger equation reduces to the one-dimensional equation by the substitution of the radial wave function

$$R_{nl}(r) = u_{nl}(r)/r.$$

On the other hand the same differential equation can be presented as an equation determining the eigenvalues of the nuclear charge for the arbitrary value of the energy E :

$$\left[-\frac{d^2}{2dr^2} + \frac{l(l+1)}{2r^2} - E \right] u_{nl}(r) = \frac{Z_n}{r} u_{nl}(r). \qquad (3.20)$$

Here the notation

$$Z_n = n\sqrt{-2E}.$$

is introduced. We solve Eq.(3.20) taking into account the boundary conditions that the solution should be finite at $r = 0$ and $r \rightarrow \infty$. Let us substitute the independent variable

$$r \equiv \frac{x}{2\sqrt{-2E}}.$$

Then the Whittaker equation follows from Eq.(3.20) (see also Eq.(3.16)):

$$\frac{d^2 u_{nl}(x)}{dx^2} + \left[\frac{\nu}{x} + \frac{1/4 - \mu^2}{x^2} - \frac{1}{4} \right] u_{nl}(x) = 0,$$ (3.21)

Here the notation

$$\nu \equiv \frac{Z_n}{\sqrt{-2E}}, \qquad \mu = l + \frac{1}{2}$$

is introduced which is analogously to the notation in the previous section. The solution of Eq.(3.21) which is finite at the origin is one of the Whittaker functions (see previous section)

$$u_{nl}(x) = \text{const} \cdot M_{\nu,\mu}(x).$$

Let us note that this Whittaker function can be expressed via the degenerate hypergeometric function

$$M_{\nu,\mu}(x) = x^{\mu + 1/2} \exp(-x/2) \cdot F(\mu - \nu + 1/2, 2\mu + 1, x).$$

The degenerate hypergeometric function is represented by a Taylor series in x:

$$F(a, b, x) = 1 + \frac{ax}{1! \, b} + \frac{a(a+1)x^2}{2! \, b(b+1)} + \dots.$$ (3.22)

In order for this solution to be finite as $x \to \infty$, the Taylor series must contain a finite number of terms. In the opposite case the degenerate hypergeometric function is of the form $\exp(x)$, $x \to \infty$. Hence the Whittaker function is of the form $\exp(x/2)$, $x \to \infty$, i.e. it is infinite.

A finite number of terms is present under the condition that the parameter a in Eq.(3.22) is any negative integer, or zero. Thus, we must require that

$$\mu - v + \frac{1}{2} = -n_r , \quad n_r = 0,1,2,3,\ldots .$$

The quantity n_r is the so-called *radial quantum number*, while the quantity

$$n = v = n_r + \mu + \frac{1}{2} = n_r + l + 1$$

is the so-called *principal quantum number*.

Thus the eigenvalue of the charge is equal to

$$Z_n = n\sqrt{-2E} ,$$

and the corresponding eigenfunction is of the form

$$u_{nl}(r) = \text{const} \cdot M_{n,\mu}\left(2\sqrt{-2Er}\right).$$

In order to derive the value of the constant in this expression we consider the normalized condition for the eigenfunctions which belong to different eigenvalues at the same value of energy E.

Multiplying Eq.(3.20) on the left by one of the eigenfunctions and integrating it over the radial variable r, and then subtracting the analogous equation for the second eigenfunction, multiplying on the left by the first eigenfunction, we obtain

$$\int_0^\infty u_{nl}(r)u_{n'l}(r)\frac{dr}{r} = \delta_{nn'}. \tag{3.23}$$

Substituting the above expression for the eigenfunction into Eq.(3.23), we can derive the constant. The final expression for the normalized eigenfunction (called the *Sturmian function*) can be written in the form

$$u_{nl}(r) = \frac{1}{(2l+1)!}\sqrt{\frac{\Gamma(n+l+1)}{\Gamma(n-l)}}M_{n,l+1/2}\left(2\sqrt{-2Er}\right). \tag{3.24}$$

The condition of completeness of the Sturmian function is of the simple form

$$\sum_{n=l+1}^{\infty} u_{nl}(r)u_{nl}(r') = r\delta(r - r').\qquad(3.25)$$

This can be proved by the following considerations. Let us multiply Eq.(3.25) by the function

$$\frac{1}{r}u_{n'l}(r).$$

Then let us integrate the resulting expression over the variable r. Using the normalization condition (3.23) we find

$$\sum_{n=l+1}^{\infty} \delta_{nn'}u_{nl}(r') = u_{n'l}(r'),$$

i.e. we obtain the required identity.

It should be noted that if the energy is

$$E = E_n = -\frac{Z^2}{2n^2},$$

then the eigenfunction (3.24) coincides with the hydrogen-like wave eigenfunction.

3.5 STURMIAN EXPANSION OF GREEN'S FUNCTIONS

Equation (3.15) for the radial Green function of a hydrogen atom can be simply generalized to the case of hydrogen-like atom with nuclear charge Z:

$$\left[-\frac{1}{2}\frac{d^2}{dr^2} + \frac{l(l+1)}{2r^2} - \frac{Z}{r} - E \right] G_{El}(r,r') = \delta(r - r').$$ (3.26)

Let us now write the expansion of this Green's function with respect to the Sturmian eigen-functions that were introduced in the previous section (see Eq. (3.24)):

$$G_{El}(r,r') = \sum_{n=l+1}^{\infty} \frac{u_{nl}(r)u_{nl}(r')}{Z_n - Z}.$$ (3.27)

Here the notation $Z_n = n(2E)^{1/2}$ is used (see previous section). In order to prove Eq.(3.27) let us substitute it into Eq.(3.26) and use Eq.(3.20) for the Sturmian functions. We obtain:

$$\sum_{n=l+1}^{\infty} \left(\frac{Z_n}{r} - \frac{Z}{r} \right) \frac{u_{nl}(r)u_{nl}(r')}{Z_n - Z} = \delta(r - r').$$

It is seen that this relation coincides with the condition of completeness (3.25) of Sturmian functions. The expansion (3.27) has some advantages with respect to the general expansion (3.2) since here we have only a summation over the states of the discrete spectrum while the sum in Eq.(3.2) is taken over the states of the discrete spectrum and over the continuum states. Integration over continuum states is very cumbersome if it is made on the computer. It should be noted that discrete and continuum states produce the comparable contributions into the sum of the type Eq.(3.2), as a rule.

The next advantage of the expansion of Green's functions over Sturmian functions is that the dependence on the independent variables r and r' is factorized unlike the closed form (3.19). In the last case the numerical calculations are very cumbersome since it is necessary to divide the integration plane into regions where one radial variable is larger or smaller than the other. Therefore the Sturmian expansion of the Green's functions is very useful, for example, in the calculation of matrix elements appearing in high-order perturbation theory: multiple integrals reduce to a product of single integrals. This fact decreases the execution time of computer calculation significantly.

It follows from Eq.(3.27) that the Green's function has single poles at the energies which correspond to discrete eigenvalues, i.e.

$$E = E_n = -\frac{Z^2}{2n^2}.$$

The expressions for Green's functions can be simply generalized to the case when the hydrogen atom is perturbed by a monochromatic electromagnetic field. If the frequency of this field is equal to ω, then Eq.(3.2) is generalized by the substitution $E \to E + \omega$. This generalization is also valid for other representations of the Green's function, in particular, for Eq.(3.27). This change allows us to obtain simple analytical expressions, for example, for two-photon matrix elements of atomic transitions. An analogous procedure allows us to derive multiphoton matrix elements.

PROBLEMS

Problem 1. Starting from Eq.(3.19) obtain the integral representation of the radial Coulomb Green's function:

$$G_{El}(r, r')$$

$$= 2\sqrt{r \cdot r'} \int_0^\infty \exp\left\{\left[-\frac{r + r'}{\nu}\right]\cosh t\right\}\left[\coth\frac{t}{2}\right]^{2\nu} I_{2l+1}\left\{\frac{2\sqrt{r \cdot r'}}{\nu}\sinh t\right\}dt$$

Here $I_{2l+1}(x)$ is the so-called *modified Bessel function*.

Problem 2. Obtain an expression for the solution of the nonhomogeneous differential equation

$$H\psi(\mathbf{r}) - E\psi(\mathbf{r}) = f(\mathbf{r})$$

(H is the differential operator) via the Green's function, Eq.(3.2), in the form

$$\psi(\mathbf{r}) = \int G_E(\mathbf{r}, \mathbf{r}')f(\mathbf{r}')d\mathbf{r}'.$$

Problem 3. Calculate the Green's function for the wave equation

$$\left(\Delta - \frac{1}{c^2}\frac{\partial^2}{\partial t^2}\right)G(\mathbf{r} - \mathbf{r}') = \delta(\mathbf{r} - \mathbf{r}')\delta(t - t')$$

(Δ is the Laplace operator acting on the coordinate $\mathbf{r}$).
Answer:

$$G(\mathbf{r} - \mathbf{r}', t - t') = -\frac{c}{4\pi|\mathbf{r} - \mathbf{r}'|}\delta\left\{|\mathbf{r} - \mathbf{r}'| - c(t - t')\right\}, \quad t > t';$$

$$G(\mathbf{r} - \mathbf{r}', t - t') = 0, \qquad t < t'.$$

Problem 4. Using the solutions to Problems 2 and 3 obtain the solution of the nonhomogeneous wave equation

$$\left(\Delta - \frac{1}{c^2}\frac{\partial^2}{\partial t^2}\right)\Psi(\mathbf{r}, t) = f(\mathbf{r}, t).$$

Answer:

$$\Psi(\mathbf{r}, t) = -\frac{1}{4\pi}\int \frac{f\left[\mathbf{r}', t - \dfrac{|\mathbf{r} - \mathbf{r}'|}{c}\right]}{|\mathbf{r} - \mathbf{r}'|}d\mathbf{r}'.$$

In electrodynamics this solution determines the so-called *retarding potential.* This name is explained by the retarding time in the integrand. The difference in time is equal to the time of propagation of a signal from the source up to the observation point travelling at the light speed c.

Problem 5. Solve the previous problem for the case of unit point charge, moving according to the given classical trajectory $\mathbf{R}(t)$, i.e.

$$f(\mathbf{r}, t) = -4\pi\delta(\mathbf{r} - \mathbf{R}(t))..$$

Answer:

$$\Psi(\mathbf{r}, t) = \cfrac{1}{\left|\mathbf{r} - \mathbf{R}(t')\right| - \cfrac{1}{c}(\mathbf{r} - \mathbf{R}(t'))\cfrac{d\mathbf{R}(t')}{dt}},$$

where the retarding time t' is determined from the implicit equation

$$t' = t - \frac{1}{c}\left|\mathbf{r} - \mathbf{R}(t')\right|..$$

This is so-called Lienard – Wiechert potential.

Problem 6. Solve the equation for the Green's function describing the Coulomb electrostatic potential of a unit point charge in cylindrical coordinates:

$$\Delta G(\mathbf{r}, \mathbf{r}') = -\frac{4\pi}{\rho}\delta(\rho - \rho')\delta(\varphi - \varphi')\delta(z - z').$$

Here the delta function is expressed in cylindrical coordinates.
Answer:

$$G(\mathbf{r}, \mathbf{r}')$$

$$= \frac{1}{2\pi^2}\sum_{m=-\infty}^{\infty}\exp[im(\varphi - \varphi')\int_0^\infty dk\,\cos[k(z - z')g_m(k, \rho, \rho').$$

The radial Green's function is of the form

$$g_m(k, \rho, \rho') = 4\pi\,I_m(k\rho_<)K_m(k\rho_>).$$

Here I_m and K_m are the *modified Bessel functions* (Bessel functions of pure imaginary argument). They satisfy the differential equation

$$\frac{d^2 R}{d\rho^2} + \frac{1}{\rho}\frac{dR}{d\rho} - \left(k^2 + \frac{m^2}{\rho^2}\right)R = 0.$$

Problem 7. The geometry of a two-dimensional electrostatic potential is defined in polar coordinates by the grounded surfaces $\varphi = 0$, $\varphi = \alpha$, $\rho = a$. Using separation of variables in polar coordinates, show that the Green's function (the electrostatic potential at the point (ρ, φ) produced by a unit line charge at the point (ρ', φ')) can be written as

$$G(\rho, \varphi; \rho', \varphi')$$

$$= \sum_{m=1}^{\infty} \frac{4}{m}\, \rho_<^{\pi m/\alpha} \left[\frac{1}{\rho_>^{\pi m/\alpha}} - \frac{\rho_>^{\pi m/\alpha}}{a^{2\pi m/\alpha}}\right] \sin\left(\frac{\pi m \varphi}{\alpha}\right) \sin\left(\frac{\pi m \varphi'}{\alpha}\right)$$

Problem 8. Calculate the Green's function (electrostatic potential) for the following electrostatic problem: grounded sphere of radius a; the potential at the location $\mathbf{r}$ is produced by unit point charge at the location $\mathbf{r}'$.
Answer:

$$G(\mathbf{r}, \mathbf{r}') = \frac{1}{|\mathbf{r} - \mathbf{r}'|} - \frac{a}{r'\left|\mathbf{r} - \left(\dfrac{a}{r'}\right)^2 \mathbf{r}'\right|}.$$

Chapter 4

The Whittaker Method

4.1 FLOQUET SOLUTIONS

The Whittaker method is a powerful approximation for the perturbative consideration of various differential equations containing *resonance increasing* solutions. We consider this method for the example of linear ordinary differential equations of the second order with periodic coefficients.

In linear vibrational systems the parameters of oscillations, for example, the frequency of oscillations, or the mass of a particle, can periodically vary with time. In such cases we talk about parametric excitation of the vibrational system. The simplest linear ordinary differential equation which describes this process is the so-called *Mathieu equation*

$$\frac{d^2 y(t)}{dt^2} + [\Omega^2 + k \cos \omega t] y(t) = 0. \tag{4.1}$$

Here Ω and ω are the frequency of vibrations of the system, and the frequency of parametric excitation, respectively. The parameter k is the modulation amplitude. We assume that this amplitude is very small, i.e. $k << \Omega^2$. However, the smallness of the modulation amplitude does not imply the smallness of the perturbative terms in the solution of Eq.(4.1).

Let $y_1(t)$ be some solution of Eq.(4.1) with the simple initial conditions

$$y_1(0) = 1, \quad y_1'(0) = 0.$$

Then the expression $y_1(t + 2\pi/\omega)$ is also a solution of Eq.(4.1) . Hence, we can expand it with respect to the system of two linear independent solutions:

$$y_1\left(t + \frac{2\pi}{\omega}\right) = a_{11}y_1(t) + a_{12}y_2(t). \tag{4.2}$$

Here $y_2(t)$ is the solution of Eq. (4.1) which is linearly independent with respect to the first solution $y_1(t)$. The quantities a_{11}, a_{22} are expansion coefficients. The initial conditions for the second solution are of the form

$$y_2(0) = 0, \quad y_2'(0) = 1.$$

Thus the Wronskian of the system of solutions $[y_1, y_2]$ is equal to unity. Analogously we determine

$$y_2\left(t + \frac{2\pi}{\omega}\right) = a_{21}y_1(t) + a_{22}y_2(t). \tag{4.3}$$

Let us now calculate the eigenvalues of the transformation matrix after the time period $2\pi/\omega$:

$$A = \begin{pmatrix} a_{11} & a_{12} \\ a_{21} & a_{22} \end{pmatrix} \tag{4.4}$$

They can be derived from the determinant equation

$$\begin{vmatrix} a_{11} - \lambda & a_{12} \\ a_{21} & a_{22} - \lambda \end{vmatrix} = 0. \tag{4.5}$$

Its solution is of the form

$$\lambda_{1,2} = \alpha \pm \sqrt{\alpha^2 - 1}. \tag{4.6}$$

Here the notation

$$\alpha = \frac{a_{11} + a_{22}}{2} \tag{4.7}$$

is introduced. We take into account that according to Eqs. (4.2) and (4.3) the determinant of the matrix A is equal to unity:

$$\det A = a_{11}a_{22} - a_{12}a_{21}$$

$$= y_1\left(\frac{2\pi}{\omega}\right)y_2'\left(\frac{2\pi}{\omega}\right) - y_1'\left(\frac{2\pi}{\omega}\right)y_2\left(\frac{2\pi}{\omega}\right) = 1. \tag{4.8}$$

This follows from the fact that the Wronskian of the system $[y_1, y_2]$ is constant and that the value of this constant is equal to 1 due to the initial conditions for the eigenfunctions y_1, y_2.

According to the rules of linear algebra there is a linear transformation from the system $[y_1, y_2]$ to the new system of eigenfunctions $[x_1, x_2]$ so that the new eigenfunctions transform into themselves:

$$x_1\left(t + \frac{2\pi}{\omega}\right) = \lambda_1 x_1(t); \quad x_2\left(t + \frac{2\pi}{\omega}\right) = \lambda_2 x_2(t). \tag{4.9}$$

The solutions x_1, x_2 are the so-called *Floquet solutions.*

4.2 FLOQUET EIGENVALUES

4.2.1. *The case of* $|\alpha| > 1$.

It follows from Eq.(4.9) that

$$x_1\left(t + \frac{2\pi n}{\omega}\right) = \lambda_1^n x_1(t). \tag{4.10}$$

According to Eq.(4.6) the eigenvalues λ_1, λ_2 are real under the condition $|\alpha| > 1$. If $|\lambda_1| < 1$, then according to Eq.(4.10) we have $x_1(\infty) \to 0$. Thus, in this case the solution $x_1(t)$ is *stable*. According to Eq. (4.6) the product of the eigenvalues is equal to 1. Hence, $|\lambda_2| > 1$, $x_2(\infty) \to \infty$. Thus, the solution $x_2(t)$ is *unstable*. In the opposite case $|\lambda_1| > 1$, $|\lambda_2| < 1$ we find that the solution $x_2(t)$ is stable while the solution $x_1(t)$ is unstable.

Finally, if $\lambda_1 = \lambda_2 = 1$, then according to Eq.(4.9) the eigenfunctions $x_1(t)$, $x_2(t)$ are periodic with period $2\pi/\omega$. If $\lambda_1 = \lambda_2 = -1$, then according to Eq.(4.9) the eigenfunctions $x_1(t)$, $x_2(t)$ are periodic with period $4\pi/\omega$. The values of λ_1, $\lambda_2 = \pm 1$ are called *transition values* since they separate stable solutions from unstable solutions.

Let us consider the function

$$u_1(t) = \lambda_1^{(-\omega t / 2\pi)} x_1(t). \qquad (4.11)$$

It follows from Eqs.(4.9) and (4.11) that

$$u_1\left(t + \frac{2\pi}{\omega}\right) = u_1(t), \qquad (4.12)$$

i.e. $u_1(t)$ is a periodic function with period $2\pi/\omega$.

Thus, the solutions $x_1(t)$, $x_2(t)$ can be represented in the so-called *Floquet form* :

$$x_{1,2}(t) = \exp(\gamma_{1,2}t) \cdot u_{1,2}(t) \qquad (4.13)$$

where the quantities

$$\gamma_{1,2} \equiv \frac{\omega}{2\pi} \ln \lambda_{1,2} \qquad (4.14)$$

are the so-called *characteristic exponents*. The functions $u_{1,2}(t)$ are periodic with period $2\pi/\omega$. The positive value of the characteristic exponent $\gamma_1 > 0$ corresponds to the eigenvalue $|\lambda_1| > 1$. Hence, the positive characteristic exponent determines the unstable solution. Similarly, the negative characteristic exponent determines the stable solution.

4.2.2. *The case of* $|\alpha| < 1$.

Now we consider the opposite case $|\alpha| < 1$. Then according to Eq.(4.6) we have

$$\sqrt{\alpha^2 - 1} = i\sqrt{1 - \alpha^2}.$$

Hence, $\lambda_2 = \lambda_1^*$ so that $|\lambda_1| = |\lambda_2| = 1$. We can conclude that $\mathrm{Re}\,\gamma_1 = \mathrm{Re}\,\gamma_2 = 0$ and according to Eq.(4.13) the solutions $x_1(t)$, $x_2(t)$ are confined though they can be both periodic and non periodic. More exactly, these solutions have two periods:

$$T_1 = 2\pi / \omega, \qquad T_2 = 2\pi / \mathrm{Im}\,\gamma_{1,2}.$$

Finally let us consider the case $|\alpha| = 1$. If $\alpha = 1$, then the transition values are of the form $\lambda_1 = \lambda_2 = 1$, $\gamma_1 = \gamma_2 = 0$ so that the functions $x_{1,2}(t)$ are periodic with period $T = 2\pi/\omega$. If now $\alpha = -1$ then $\lambda_1 = \lambda_2 = 1$, $\gamma_1 = \gamma_2 = i\omega/2$ and according to Eq.(4.12) the period of the functions $x_{1,2}(t)$ is equal to $T = 4\pi/\omega$.

Let us obtain now the solution of the Mathieu equation (4.1) for $k << \Omega^2$ for transition values (these are so called *transition solutions*).

4.3 TRANSITION SOLUTIONS

We consider in this section the transition solutions of Eq.(4.1) which are periodic functions. Further we change the independent variable: $\omega t \to 2t$. Then Eq.(4.1) is transformed to the traditional form of the Mathieu equation:

$$\frac{d^2 y(t)}{dt^2} + (d + \varepsilon \cos 2t)y(t) = 0. \qquad (4.15)$$

Here the notation

is introduced. The new parameters are dimensionless quantities. It should be noted that the parameter ε can be both positive and negative.

The solution of Eq.(4.15) is presented in the iterative form:

$$y(t) = y_0(t) + \varepsilon y_1(t) + \varepsilon^2 y_2(t) + ...;$$

$$d = n^2 + \varepsilon d_1 + \varepsilon^2 d_2 + ...$$

$$(4.16)$$

Here n is still an arbitrary number. Substituting Eq.(4.16) into Eq.(4.15) we obtain the following equation for the zero iteration

$$\frac{d^2 y_0(t)}{dt^2} + n^2 y_0(t) = 0. \qquad (4.17)$$

The solution of this equation is of the simple form

$$y_0(t) = a \cos nt + b \sin nt. \qquad (4.18)$$

It is periodic with period $T = 2\pi/n$.

The first iteration function $y_1(t)$ satisfies the equation:

$$\frac{d^2 y_1(t)}{dt^2} + n^2 y_1(t) = -d_1 y_0(t) - y_0(t) \cos 2t. \qquad (4.19)$$

Substituting Eq.(4.18) into Eq.(4.19) we obtain the following nonuniform differential equation for the first iteration $y_1(t)$:

$$\frac{d^2 y_1}{dt^2} + n^2 y_1 = -[a \cos(nt) + b \sin(nt)](\cos 2t + d_1). \qquad (4.20)$$

The function $y_1(t)$ is periodic with the same period $2\pi/\omega$ only at the definite values of n and d_1 in the right side of Eq.(4.20).

Let us choose, for example, $n = 1$. Then Eq.(4.20) takes the form

$$\frac{d^2 y_1}{dt^2} + y_1 \tag{4.21}$$

$$= -d_1 \, (a \cos t + b \sin t) - \frac{a}{2} \, (\cos t + \cos 3t) + \frac{b}{2} \, (\sin t - \sin 3t).$$

In order to exclude the secular terms on the right side of Eq.(4.21) which are proportional to $\cos t$ and $\sin t$ (these terms produce non-periodic solutions) it is necessary that the two conditions satisfy

$$\left(d_1 + \frac{1}{2} \right) a = 0 \tag{4.22}$$

and

$$\left(d_1 - \frac{1}{2} \right) b = 0 \tag{4.23}$$

simultaneously. Then the solution $y_1(t)$ is periodic with period 2π (terms on the right side of Eq.(4.21) which are proportional to $\sin 3t$ and $\cos 3t$ are also periodic functions).

Thus we conclude that the value of n is an integer for transition solutions.

4.4 TRANSITION SOLUTION WITH $n = 1$

We consider as an example the value of $n = 1$ and derive the corresponding transition (periodic) solution. It follows from Eqs. (4.22) and (4.23) that $d_1 = 1/2$, $a = 0$ for the first transition solution, while $d_1 = -1/2$, $b = 0$ for the second solution. According to Eq. (4.21) we obtain the following equation for the first transition solution:

$$\frac{d^2 y_1}{dt^2} + y_1 = -\frac{b}{2} \sin 3t. \tag{4.24}$$

The solution of this equation is of a simple form. The sum of the zero and first iterations is

$$y(t) = b\left(\sin t + \frac{\varepsilon}{16} \sin 3t + \dots \right) \qquad (4.25)$$

The second transition solution is obtained analogously:

$$y(t) = a\left(\cos t + \frac{\varepsilon}{16} \cos 3t + \dots \right) \qquad (4.26)$$

According to Eq.(4.16) the value of the parameter d for the two transition solutions is equal to

$$d = 1 \pm \frac{\varepsilon}{2} + \dots \qquad (4.27)$$

respectively.

The period of both solutions (4.25) and (4.26) is equal to 2π. If we return to the old independent variable t then we find that the period of these solutions is equal to $4\pi/\omega$ which corresponds to the eigenvalues $\lambda_1 = \lambda_2 = = -1$.

4.5 TRANSITION SOLUTION WITH $n = 0$

Let us choose the case $n = 0$ as the second example of transition solutions. It follows from Eq. (4.19) that for the zero iteration we have the simple result $y_0(t) = a$. Then we obtain from Eq. (4.20) the equation for the first iteration:

$$\frac{d^2 y_1}{dt^2} = -a(d_1 + \cos 2t). \qquad (4.28)$$

In order for the solution of this equation to be a periodic function it is necessary that the condition $d_1 = 0$ should be fulfilled. Hence,

$$y_1(t) = \frac{a}{4} \cos 2t \; . \tag{4.29}$$

According to Eq.(4.15) we find the equation for the second iteration

$$\frac{d^2 y_2}{dt^2} = d_2 y_0 + y_1 \cos 2t. \tag{4.30}$$

Substituting Eq.(4.29) into the right side of Eq.(4.30) we rewrite Eq.(4.30) in the form

$$\frac{d^2 y_2}{dt^2} = -\left(d_2 + \frac{1}{8} \right) a - \frac{a}{8} \cos 4t. \tag{4.31}$$

In order for the solution of this equation to be a periodic function it is necessary that $d_2 = -1/8$. Then we obtain one periodic solution with $n = 0$:

$$y(t) = a \left(1 + \frac{\varepsilon}{4} \cos 2t + \frac{\varepsilon^2}{128} \cos 4t + \ldots \right) \tag{4.32}$$

Its period is equal to π. This period is equal to $2\pi/\omega$ in terms of the old independent variable t which corresponds to the eigenvalue $\lambda = 1$.

The quantity d for this transition solution is of the form

$$d = -\frac{\varepsilon^2}{8} + \ldots \tag{4.33}$$

The dependencies of the type of Eq.(4.27), (4.33) are called *transition curves*. They are depicted in Fig. 3 for the arbitrary values of ε as functions of ε.

Calculations for large values of ε can be made only numerically. The transition curves separate regions where the solutions are stable and unstable (the latter are shaded).

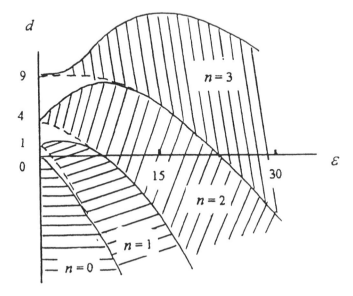

Figure 3. The transition curves for the Mathieu equation. The regions are shaded where the solution are unstable. Other regions correspond to stable solutions. The transition curves with $n = 0$, 1, 2, and 3 are shown.

4.6 THE WHITTAKER METHOD

Let us now obtain the solutions of the Mathieu equation near the transition curves when $\varepsilon \ll 1$. We choose the case $n = 1$ as a typical example and from the two transition curves with $n = 1$ we choose the vicinity of the transition curve $d = 1 - \varepsilon/2$ (see Fig. 3 and Eq.(4.27)). According to Eq.(4.13) we choose the solution of Eq.(4.15) in the Floquet form

$$y(t) = \exp(\gamma t) \cdot u(t) \tag{4.34}$$

where γ is the characteristic exponent and $u(t)$ is a periodic function. Substituting Eq.(4.34) into Eq.(4.15) we obtain the differential equation for the periodic function $u(t)$:

$$\frac{d^2 u}{dt^2} + 2\gamma \frac{du}{dt} + \left(\gamma^2 + d + \varepsilon \cos 2t \right) u(t) = 0. \tag{4.35}$$

The periodic solution on the transition curve is given by Eq.(4.26). The solution of Eq.(4.35) is presented in the iterated form:

$$u(t) = u_1(t) + \varepsilon u_2(t) + \dots$$

$$\gamma = \varepsilon \gamma_1 + \varepsilon^2 \gamma_2 + \dots \tag{4.36}$$

$$d = 1 + \varepsilon d_1 + \dots.$$

The equation for zero iteration is of a simple form:

$$\frac{d^2 u_0(t)}{dt^2} + u_0(t) = 0. \tag{4.37}$$

Its solution is

$$u_0(t) = a \cos t + b \sin t. \tag{4.38}$$

The equation for the first iteration $u_1(t)$ is of the form

$$\frac{d^2 u_1(t)}{dt^2} + u_1(t) = -2\gamma_1 \frac{du_0(t)}{dt} - d_1 u_0(t) - u_0(t) \cos 2t. \tag{4.39}$$

Substituting Eq.(4.38) into Eq.(4.39) we obtain the differential equation:

$$\frac{d^2 u_1(t)}{dt^2} + u_1(t) = -2\gamma_1 (-a \sin t + b \cos t) - d_1 (a \cos t + b \sin t)$$
$$\tag{4.40}$$

$$-\frac{1}{2} a(\cos t + \cos 3t) + \frac{1}{2} b(\sin t - \sin 3t).$$

In order for the solution $u_1(t)$ to be periodic (according to Floquet's theorem) it is necessary that the secular terms on the right side of Eq.(4.40) that produce increasing nonperiodic solutions should be equal to zero.

Thus, two conditions should be fulfilled simultaneously:

$$2\gamma_1 a - d_1 b + \frac{1}{2} b = 0$$

$$- 2\gamma_1 b - d_1 a - \frac{1}{2} a = 0.$$

(4.41)

From the solution of the system of two equations (4.41) we find the characteristic exponent

$$\gamma_1 = \pm \frac{1}{2} \sqrt{\frac{1}{4} - d_1^2},$$

(4.42)

as well as the relation between the coefficients a and b :

$$b = \frac{2\gamma_1}{d_1 - \frac{1}{2}} a = \mp \left(\frac{1 + 2d_1}{1 - 2d_1} \right)^{1/2} a.$$

(4.43)

Thus the zero iteration of the solution near the transition curve $d = 1 - \varepsilon/2$ is of the form

$$y(t) = a_1 \left[\cos t - \left(\frac{1 + 2d_1}{1 - 2d_1} \right)^{1/2} \sin t \right] \exp\left\{ \frac{\varepsilon t}{2} \sqrt{\frac{1}{4} - d_1^2} \right\}$$

$$+ a_2 \left[\cos t + \left(\frac{1 + 2d_1}{1 - 2d_1} \right)^{1/2} \sin t \right] \exp\left\{ -\frac{\varepsilon t}{2} \sqrt{\frac{1}{4} - d_1^2} \right\}.$$

(4.44)

Here a_1, a_2 are integration constants; the quantity $d_1 = (d - 1)/\varepsilon$ is approximately $-1/2$.

If $d > 1 - \varepsilon/2$ (this is the region above the corresponding transition curve in Fig. 3) then $d_1 > -1/2$ so that the quantity

$$\sqrt{1/4 - d_1^2}$$

is real. In this case the solution Eq.(4.44) increases as the variable t increases, i.e. it is unstable.

Below the transition curve we have $d_1 < -1/2$ so that the quantity

$$\sqrt{1/4 - d_1^2}$$

is imaginary. Hence, the solution Eq.(4.44) is a two-periodic one, i.e. it is stable. Its periods are equal to

$$T_1 = 2\pi; \qquad T_2 = \frac{4\pi}{\varepsilon\sqrt{d_1^2 - 1/4}}. \qquad (4.45)$$

In conclusion let us note that we suppose that $\varepsilon > 0$ in all the above considerations. It is seen from Fig. 3 that the transition curves are symmetrical with respect to the vertical axis, so all the above results can be transferred into the region of negative values of the parameter ε.

4.7 THE METHOD OF MANY SCALES FOR NONLINEAR EQUATIONS

Differential equations in physical problems often contain small *nonlinear* terms. Direct expansion of the solution over the small parameter ε is incorrect, as a rule, since the product of this parameter with the independent variable, x , can be very large, i.e., the inequalty $\varepsilon x \gg 1$ violates the applicability of the standard expansion in the iteration procedure. So-called divergent *secular terms* appear in direct iterations.

One of the powerful methods which allows us to avoid secular terms and to find correct iterative series over the small parameter ε is the method of many scales. In its simplest form this method consists of the statement that the solution of a nonlinear differential equation is determined by two different scales: $x_1 = x$ and $x_2 = \varepsilon x \ll x_1$. Of course, if the value of the independent variable x is so large that $\varepsilon^2 x > 1$, then we should introduce the third scale $x_3 = \varepsilon^2 x$, and so on.

The quantities $x_1 = x$, $x_2 = \varepsilon x$, $x_3 = \varepsilon^2 x$, and so on, are considered in this approach as *independent* variables. Then the ordinary differential equation takes the form of a *partial* differential equation. The number of integration constants in the iteration procedure for the partial differential equation is more than that for the ordinary differential equation. However, there is the possibility of choosing these constants so that all secular terms vanish. Then the obtained solution is valid for any values of the independent variable x.

As always, we consider a simple physical example, instead of the general description of the method. Let us consider the so-called *Van der Pol equation*

$$\frac{d^2 y}{dx^2} + y = \varepsilon \frac{dy}{dx}\left(1 - y^2\right) \qquad \varepsilon > 0; \qquad \varepsilon \ll 1. \qquad (4.46)$$

This equation describes oscillations with weak friction, but the friction can be positive, or negative, depending on the amplitude of the oscillations.

We introduce the independent variables

$$x_1 \equiv x, \qquad x_2 \equiv \varepsilon x, \qquad x_3 \equiv \varepsilon^2 x... \qquad (4.47)$$

The first derivative takes the form

$$\frac{d}{dx} = \frac{\partial}{\partial x_1} + \varepsilon \frac{\partial}{\partial x_2} + ...$$

The second derivative takes the form

$$\frac{d^2}{dx^2} = \frac{\partial^2}{\partial x_1^2} + 2\varepsilon \frac{\partial^2}{\partial x_1 \partial x_2} + ...$$

Substituting this expression into Eq. (4.47), we obtain

$$\frac{\partial^2 y}{\partial x_1^2} + 2\varepsilon \frac{\partial^2 y}{\partial x_1 \partial x_2} + y = \varepsilon \frac{\partial y}{\partial x_1}\left(1 - y^2\right) \qquad (4.48)$$

We restrict ourselves to terms which are linear in ε.

The solution of Eq. (4.48) is presented in the iterative form

$$y = y_0 + \varepsilon y_1 + \dots$$

Substituting this expansion in Eq. (4.48), we find first the equation for the zero iteration y_0:

$$\frac{\partial^2 y_0}{\partial x_1^2} + y_0 = 0.$$

Its solution is of the simple form

$$y_0(x_1, x_2) = A(x_2)\exp(ix_1) + A^*(x_2)\exp(-ix_1). \qquad (4.49)$$

The complex quantity $A(x_2)$ does not depend on the variable x_1.

Further, from Eq. (4.48) we obtain the equation for the first iteration y_1:

$$\frac{\partial^2 y_1}{\partial x_1^2} + 2\frac{\partial^2 y_0}{\partial x_1 \partial x_2} + y_1 = \frac{\partial y_0}{\partial x_1}\left(1 - y_0^2\right) \qquad (4.50)$$

Substituting Eq. (4.49) into Eq. (4.50), we find the inhomogeneous linear differential equation for the first iteration:

$$\frac{\partial^2 y_1}{\partial x_1^2} + y_1 = -2i\frac{dA(x_2)}{dx_2}\exp(ix_1) + 2i\frac{dA^*(x_2)}{dx_2}\exp(-ix_1)$$

$$+ iA \exp(ix_1) - iA^* \exp(-ix_1) - iA^3 \exp(3ix_1)$$

$$+ iA^{*3} \exp(-3ix_1) - iA^2 A^* \exp(ix_1) + iA^{*2} A \exp(-ix_1).$$

$$(4.51)$$

It is seen that the inhomogeneous part of this equation is the solution of the corresponding homogeneous differential equation for the function y_1. This part produces a secular term which increases with a rise in x_1. In order to prevent such unlimited increasing, we should choose the condition that the secular terms are absent:

$$2 \frac{dA(x_2)}{dx_2} + A^2 A^* - A = 0. \qquad (4.52)$$

In order to solve this equation, we present the quantity A in the standard form of the complex number $A = a \exp(ib)$. Here the amplitude a and the phase b are real functions of the variable x_2. Then zero iteration of the solution (4.49) takes the simple form

$$y_0 = 2a \cos(x_1 + b). \qquad (4.53)$$

Equation (4.52) reduces to two real equations:

$$2 \frac{da}{dx_2} = a - a^3 ; \qquad \frac{db}{dx_2} = 0. \qquad (4.54)$$

It follows from the second equation in (4.54) that $b = $ const. Integrating the first ordinary differential equation of the first order in (4.54), we obtain

$$a(x_2) = \frac{1}{\sqrt{1 + C \exp(-x_2)}}. \qquad (4.55)$$

Here C is the integration constant.

It follows from (4.53) and (4.55) that the final expression for the zero iteration of the solution is

$$y_0(x) = \frac{2\cos(x + b)}{\sqrt{1 + C\exp(-\varepsilon x)}} . \qquad (4.56)$$

The constants b and C are determined from the initial (or boundary) conditions.

If $x \to \infty$, we obtain from (4.56) that

$$y_0(x) \to 2\cos(x + b). \qquad (4.57)$$

Thus, the amplitude of oscillations in the stationary regime does not depend on the initial conditions. This is the so-called *limiting cycle* of the solution. Equation (4.56) determines the law of approach of the solution to the limiting cycle.

Equation (4.51) takes the form

$$\frac{\partial^2 y_1}{\partial x_1^2} + y_1 = -iA^3 \exp(3ix_1) + iA^{*3} \exp(-3ix_1). \qquad (4.58)$$

Its solution is

$$y_1 = \frac{i}{8} A^3 \exp(3ix_1) - \frac{i}{8} A^{*3} \exp(-3ix_1) = -\frac{a^3}{4}\sin(3x + 3b). \qquad (4.59)$$

Finally, the solution, taking into account two iterations, is of the form

$$y(x) = \frac{2}{\sqrt{1 + C\exp(-\varepsilon x)}} \left\{ \cos(x + b) - \frac{\varepsilon \sin(3x + 3b)}{8[1 + C\exp(-\varepsilon x)]} \right\}. \qquad (4.60)$$

High harmonics are the typical peculiarity of the solutions of nonlinear equations.

It is seen that when $C > 0$ the amplitude of oscillations increases with increasing x, while for $C < 0$ the amplitude of oscillations decreases with increasing x.

The usual iterative scheme is applicable under the condition $\varepsilon x \ll 1$ only, when the exponent $\exp(-\varepsilon x)$ can be expanded in Taylor series.

The Van der Pol equation appears in physical problems for oscillators with self-excitation. Negative friction at small oscillation amplitudes changes to positive friction at large oscillation amplitudes. The limiting cycle with a constant oscillation amplitude is a result of such nonlinear friction.

It should be noted that the limiting cycle is stable when $\varepsilon > 0$ only. In the case of $\varepsilon < 0$ it is unstable, i.e. according to Eq. (4.60) all solutions deplete: $y = 0$ at $x \to \infty$.

PROBLEMS

Problem 1. Find the next term in the expansion of Eq.(4.33) for the transition curve with $n = 0$.
Answer:

$$d = -\frac{\varepsilon^2}{8} + \frac{7}{128}\left(\frac{\varepsilon}{2}\right)^4.$$

Problem 2. Find the next term in the expansion of Eq.(4.27) for two transition curves with $n = 1$.

Answer:

$$d^{(1)} = 1 - \frac{\varepsilon}{2} - \frac{\varepsilon^2}{32}; \qquad d^{(2)} = 1 + \frac{\varepsilon}{2} - \frac{\varepsilon^2}{32}.$$

Problem 3. Show that at asymptotically large positive values of the parameter $\varepsilon \gg 1$ the equation of all transition curves is the same and is of the very simple form

$$d \approx -\varepsilon.$$

Analogously, at asymptotically large negative values of this parameter the equation of all transition curves is of the form

$$d \approx \varepsilon.$$

Problem 4. Consider the equation

$$\frac{d^2 u}{dt^2} + (\delta + \varepsilon \cos^3 t)u = 0.$$

Find the expansions up to second order for the first three transition curves by the Whittaker method.

Problem 5. Consider the equation

$$(1 + \varepsilon \cos 2t)\frac{d^2 u}{dt^2} + \delta u = 0.$$

Using the Whittaker method find the expansion of the solution up to second order in the vicinity of transition curves with $\delta = 0, 1, 4$.

Problem 6. Consider the equation

$$\frac{d^2 u}{dt^2} + \varepsilon \cos t \cdot u = 0, \qquad \varepsilon \ll 1.$$

Using the Whittaker method show that its solution is of the form

$$u(t) = u_0 \cos\left(\frac{\varepsilon}{\sqrt{2}}t + \varphi_0\right)$$

Problem 7. Consider the Mathieu equation

$$\frac{d^2 y}{dt^2} + (d + \varepsilon \cos 2t)y = 0, \qquad d \gg \varepsilon.$$

Show that the solution of this equation is of the form:

$$y(t) = y_0 \cos\left[\sqrt{d}\left(t + \frac{\varepsilon}{4d}\sin 2t\right) + \varphi_0\right].$$

Problem 8. Consider the equation for the linear harmonic oscillator with a frequency which adiabatically changes slowly with time:

$$\frac{d^2 y}{dt^2} + f^2(\varepsilon t)y = 0, \qquad 1 >> \varepsilon > 0.$$

Show that under this condition the so-called *Ehrenfest adiabatic invariant* is

$$J = \frac{(dy/dt)^2 + f(\varepsilon t)y^2}{f(\varepsilon t)} \approx \text{const}$$

i.e., it is approximately constant with time, though the frequencies $f(-\infty)$ and $f(\infty)$ can strongly differ from one other.

Problem 9. Consider the nonlinear differential equation

$$\frac{d^2 y}{dt^2} + y + \varepsilon y^3 = 0, \qquad 0 < \varepsilon << 1.$$

Using the method of many scales show that its approximate solution is of the form

$$y(t) = a \cos\left(\sqrt{1 + \frac{3}{4}\varepsilon a^2}\, t + \varphi_0\right)$$

Chapter 5

Intense Perturbations

5.1 EXPANSION IN THE INVERSE POWERS OF A PERTURBATION

The expansion in powers of a perturbation is well known: this is perturbation theory (time-independent or time-dependent). Modified perturbation theory taking into account the resonance terms has been used in the previous chapter (we called it *the iteration procedure*).

In this chapter we consider the expansion in the inverse powers of a perturbation which is correct, obviously, for *super-intense* perturbations. Analogously to the previous chapter, instead of the general approach we investigate here simple examples of ordinary homogeneous differential equations of second order taken from quantum mechanics. Namely, first we consider the one-dimensional Schroedinger equation which determines the energy eigenvalues and the corresponding wave eigenfunctions. The solutions for large (highly-excited) energy eigenvalues are well known: this is the so-called *WKB approximation* (see Chapter 2). On the other hand, we restrict ourselves now to the case of the ground state with the discrete energy $E_0 = E$ where the WKB approach is inapplicable. The corresponding wave eigenfunction is denoted as $\varphi(x)$.

The Schroedinger equation is of the form

$$-\frac{1}{2}\frac{d^2\varphi(x)}{dx^2} + U(x)\varphi(x) = E\varphi(x). \tag{5.1}$$

Here the mass of the particle is taken to be equal to 1, $U(x)$ is the potential energy of this particle, and E is the eigenvalue of total energy of the ground state. The Planck constant is also taken to be equal to 1. We suggest that the potential energy is negative and its minimum takes place at $x = 0$. The typical length for the variation of the potential energy is denoted as a. Let $U_0 = U(0)$ be the depth of the potential well.

We consider the potential energy in Eq.(5.1) as a perturbation. It is a super-intense perturbation if the condition

$$U_0 a^2 \gg 1 \tag{5.2}$$

is fulfilled, i.e. the potential well is very deep, or very wide (see Fig. 4). In the opposite limit $U_0 a^2 \ll 1$ the usual perturbation theory is applicable, and we do not consider it here.

$$U(x)$$

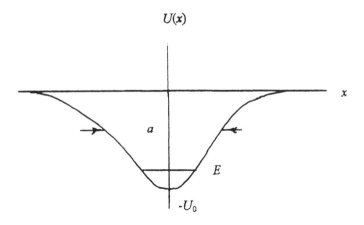

Figure 4. The potential well $U(x)$ and the the ground energy level E for Eq. (5.1).

Under the condition (5.2) we can expand the potential energy in a Taylor series with respect to the coordinate x:

$$U(x) = U_0 + \frac{1}{2}\omega^2 x^2 + \alpha x^3 + \beta x^4 + \dots \tag{5.3}$$

Here the notation

$$\omega^2 \equiv \frac{d^2 U}{dx^2}\bigg|_{x=0}, \qquad \alpha \equiv \frac{1}{6}\frac{d^3 U}{dx^3}\bigg|_{x=0}, \qquad \beta \equiv \frac{1}{24}\frac{d^4 U}{dx^4}\bigg|_{x=0}$$

(5.4)

is introduced. The first two terms in the right side of Eq.(5.3) correspond to the potential of a linear harmonic oscillator with frequency ω. The zero approximation for the energy of the ground level is given by the well known expression

$$E^{(0)} = U_0 + \frac{1}{2}\omega.$$

(5.5)

Let us calculate the next approximations for this energy. The cubic perturbation in Eq.(5.3) gives a contribution only in the second order of perturbation theory. According to this theory we have

$$E^{(2)} = \alpha^2 \sum_{m \neq 0} \frac{\left|\left\langle \varphi_0 \left| x^3 \right| \varphi_m \right\rangle\right|^2}{E^{(0)} - E_m^{(0)}}.$$

(5.6)

Here $\varphi_m(x)$ and $E_m^{(0)}$ are the unperturbed wave function of the m-th level and the corresponding energy eigenvalue, respectively.

The sum in Eq.(5.6) can be simply calculated starting with the nonzero matrix element for a transition between the oscillator states:

$$x_{m-1,m} \equiv \left\langle \varphi_{m-1} \left| x \right| \varphi_m \right\rangle = \sqrt{\frac{m}{2\omega}}.$$

(5.7)

Since for a harmonic oscillator the energy of m-th level is given by the simple expression

$$E_m^{(0)} = U_0 + \left(m + \frac{1}{2} \right)\omega$$

(5.8)

V.P. KRAINOV

we obtain from Eq.(5.6):

$$
E^{(2)} = -\alpha^2 \left\{ \frac{\left| \left(x^3 \right)_{01} \right|^2}{\omega} + \frac{\left| \left(x^3 \right)_{03} \right|^2}{3\omega} \right\}.
$$
(5.9)

The matrix elements in Eq.(5.9) can be simply derived using Eq.(5.7):

$$
(x^3)_{01} = x_{01}(x^2)_{11} = x_{01}(x_{10}x_{01} + x_{12}x_{21}) = \frac{3}{\sqrt{(2\omega)^3}}.
$$
(5.10)

Further

$$
(x^3)_{03} = x_{01}x_{12}x_{23} = \sqrt{\frac{6}{(2\omega)^3}}.
$$
(5.11)

Substituting Eqs.(5.10) and (5.11) into Eq.(5.9) we obtain the second-order correction to the energy of the ground state:

$$
E^{(2)} = -\frac{\alpha^2}{\omega^4}\left(\frac{1}{4} + \frac{9}{8} \right) = -\frac{11}{8}\frac{\alpha^2}{\omega^4}.
$$
(5.12)

The last term βx^4 in the right side of Eq.(5.3) produces the contribution to the energy already in first-order perturbation theory. This contribution is equal to

$$
E^{(1)} = \beta \left\langle \varphi_0 \left| x^4 \right| \varphi_0 \right\rangle.
$$
(5.13)

The matrix element in Eq.(5.13) is calculated analogously to Eqs.(5.10) and (5.11):

$$(x^4)_{00} = \left|(x^2)_{02}\right|^2 + \left|(x^2)_{00}\right|^2 = \left|x_{01}x_{12}\right|^2 + \left|x_{01}\right|^4 = \frac{3}{4\omega^2}.$$

(5.14)

Thus, substituting Eq.(5.14) into Eq.(5.13) we obtain

$$E^{(1)} = \frac{3\beta}{4\omega^2}.$$

(5.15)

It follows from Eqs. (5.5), (5.12) and (5.15) that the final expression for the perturbed energy of the ground state of Eq.(5.1) is

$$E = U_0 + \frac{1}{2}\omega - \frac{11}{8}\frac{\alpha^2}{\omega^4} + \frac{3\beta}{4\omega^2} + \dots$$

(5.16)

5.2 THE PARAMETER OF EXPANSION

Let us determine the parameter of expansion in the principal result (5.16). It follows from Eq.(5.4) that the estimates of the considered quantities are

$$\omega^2 \propto \frac{U_0}{a^2}, \qquad \frac{\omega}{U_0} \propto \frac{1}{\sqrt{U_0 a^2}} \ll 1.$$

(5.17)

(see the inequality (5.2)). Thus the second term in the right side of Eq.(5.16) is small compared to the first term under the condition (5.2). The estimate of the third term in the right side of Eq.(5.16) is made analogously:

$$\alpha \propto \frac{U_0}{a^3}, \qquad \frac{\alpha^2}{\omega^4 U_0} \propto \frac{1}{U_0 a^2} \ll 1.$$

(5.18)

Finally we estimate the fourth term in Eq.(5.16):

$$\beta \propto \frac{U_0}{a^4}, \qquad \frac{\beta}{\omega^2 U_0} \propto \frac{1}{U_0 a^2} \ll 1. \qquad (5.19)$$

First of all we can conclude that the terms αx^3 and βx^4 produce comparable contributions to the energy eigenvalue. Secondly, the dimensionless parameter of expansion is

$$N = \frac{1}{\sqrt{U_0 a^2}} \ll 1.$$

so that this approximation does work well for super-intense perturbations. In the next example we will see that even when this parameter is of order 1, the expansion (5.16) is also good numerically.

Sometimes this approach is called the $1/N$ *approximation*. The parameter N can be different in other problems. This is asymptotic expansion in the small quantity $1/N \ll 1$. The well-known example is the WKB approximation where this parameter is the principal quantum number of the considered quantum state (see Chapter 2).

5.3 AN EXAMPLE ON THE BOUNDARY OF APPLICABILITY

In this section we consider the solution of the Schroedinger equation for a simple potential where an analytical solution is also available:

$$U(x) = -\frac{1}{\cosh^2 x}. \qquad (5.20)$$

It is seen that no small parameter is contained here.

Our goal is to derive the energy of the ground state using the expansion (5.16). The exact analytical value of this energy is equal to $E = -0.50$. According to Eqs. (5.4) and (5.20) we obtain

$$U_0 = -1, \quad \omega^2 = 2, \quad \alpha = 0, \quad \beta = -2/3. \tag{5.21}$$

Substituting Eq.(5.21) into Eq.(5.16) we obtain $E \approx -0.54$. Thus we can conclude that approximate and exact values of the energy of the ground state $(N = 1)$ coincide with each other to within an accuracy of 8% !

It should be noted that the application of the WKB approximation (i.e. the *Bohr quantization rule*) to the ground state in the potential (5.20) gives a much worse result: $E \approx -0.42$.

For the sake of simplicity we considered above only the ground state of the quantum system, i.e. the lowest eigenvalue. It is not difficult to generalize the above derivations to the case of the excited mth level. Omitting the simple calculations we obtain

$$E_m = U_0 + \left(m + \frac{1}{2} \right) \omega - \frac{15\,\alpha^2}{4\,\omega^4} \left(m^2 + m + \frac{11}{30} \right)$$

$$+ \frac{3}{2} \frac{\beta}{\omega^2} \left(m^2 + m + \frac{1}{2} \right) + \dots \tag{5.22}$$

Equation (5.22) reduces to Eq.(5.16) for $m = 0$.

If we increase the amplitude of the perturbing potential then the quality of this approximation becomes better. For example, in the case of the potential

$$U(x) = -\frac{16}{\cosh^2 x} \tag{5.23}$$

$(N = 4)$ we obtain analogously for the energy of the ground state $(m = 0)$ $E_{exact} = -13.411$; $E_{approx} = -13.422$. The WKB approximation gives much worse value $E_{WKB} = -13.297$. Further, we find for the energy of the first excited state $E(m=1)_{exact} = -8.732$; $E(m=1)_{approx} = -8.765$. The WKB approximation gives again much worse value $E(m=1)_{WKB} = -8.640$.

5.4 LANDAU – ZENER TRANSITIONS

Now we consider another example of an intense perturbation taken from quantum mechanics. Let us investigate the dynamics of a two-level system under the action of a monochromatic super-intense field

$$V(t) = dF \sin \omega t. \qquad (5.24)$$

Here d is the dipole moment connecting both levels, F is the amplitude of the field strength of the external field, ω is the frequency of the field (for example, of the electromagnetic field of laser radiation in the dipole approximation), and t is the time.

The energies of the levels are taken as $-\Delta E/2$ (lower, 1) and $+\Delta E/2$ (upper, 2), respectively. We assume that the condition of a super-intense field

$$dF >> \Delta E \qquad (5.25)$$

is fulfilled.

Besides of this, this field is assumed to be a low-frequency field, i.e. $\omega << \Delta E$. This inequality allows us to avoid any resonance transitions in the system (the Planck constant is again chosen to be equal to 1 everywhere).

We find below that both a weak field and a super-intense field produce small perturbations in the system. Weak perturbation of the system by a super-intense field is called "stabilization" of the system. Strong perturbation is achieved only when the coupling force inside the quantum system and the force of the external field are of the same order of magnitude. This requirement is met under the obvious condition

$$dF \sin \omega t \approx dF\omega t \propto \Delta E, \qquad \omega t << 1. \qquad (5.26)$$

We have chosen here one of the periods of the monochromatic field, for the sake of simplicity.

Our goal is to solve the quantum-mechanical equations describing the dynamics of this two-level system in the field. We can extend the temporal interval up to infinity: $-\infty < t < \infty$, since, in fact, transitions take place only in the small temporal interval near $t = 0$ (the next periods of the electromagnetic field are too far from this time interval).

Thus, the Schroedinger equation for the amplitudes a_1, a_2 of the population of the levels 1 and 2, respectively, can be written in the form

$$i \frac{da_1(t)}{dt} = dF\omega t \cdot a_2(t) \exp(-i\Delta Et);$$

$$i \frac{da_2(t)}{dt} = dF\omega t \cdot a_1(t) \exp(+i\Delta Et).$$

(5.27)

The normalization condition for these amplitudes is

$$\left| a_1(t) \right|^2 + \left| a_2(t) \right|^2 = 1. \tag{5.28}$$

Further we introduce new amplitudes

$$b_1(t) \equiv a_1(t) \exp(+i\Delta Et / 2); \qquad b_2(t) \equiv a_2(t) \exp(-i\Delta Et / 2).$$

(5.29)

Substituting Eq.(5.29) into Eq.(5.27) we find the system of equations:

$$i \frac{db_1(t)}{dt} + \frac{\Delta E}{2} b_1(t) = dF\omega t \cdot b_2(t);$$

$$i \frac{db_2(t)}{dt} - \frac{\Delta E}{2} b_2(t) = dF\omega t \cdot b_1(t).$$

(5.30)

The diagonal part of the Hamiltonian represents the energies of the unperturbed quantum states while the non-diagonal part of the Hamiltonian mixes these states according to Eqs.(5.30).

Further we introduce new variables instead of b_1, b_2:

$$b_l \equiv b_1 + b_2; \qquad b_r \equiv b_1 - b_2. \tag{5.31}$$

For example, in the case of a diatomic hydrogen molecular ion the state "1" is the even ground state of electron motion, while the state "2" is the excited odd state. Then the quantity $b_l(t)$ determines the amplitude of the probability of finding an electron near the left proton while the quantity $b_r(t)$ is the amplitude of the probability of finding this electron near the right proton.

It follows from Eqs.(5.30) and (5.31) that the system of equations for the new dependent variables is

$$i \frac{db_l(t)}{dt} - dF\omega t \cdot b_l(t) = -\frac{\Delta E}{2} b_r(t);$$

(5.32)

$$i \frac{db_r(t)}{dt} + dF\omega t \cdot b_r(t) = -\frac{\Delta E}{2} b_l(t).$$

It is seen that now the quantities $\pm dF\omega t$ are the diagonal parts of the Hamiltonian while the quantity $-\Delta E/2$ mixes the unperturbed states "l" and "r".

We are interested in the probability of a transition of an electron from one state to another. In order to solve the system (5.32) we exclude from this system one dependent variable and obtain the second-order differential equation for another variable. Thus, we differentiate the first of Eqs.(5.32) over time, and substitute the derivative db_r/dt from the second of Eqs.(5.32) and the variable b_r again from the first of Eqs.(5.32). Then we find the second-order ordinary differential equations

$$\frac{d^2 b_l(t)}{dt^2} + \left[(dF\omega t)^2 + \frac{1}{4}(\Delta E)^2 + idF\omega \right] b_l(t) = 0.$$

(5.33)

$$\frac{d^2 b_r(t)}{dt^2} + \left[(dF\omega t)^2 + \frac{1}{4}(\Delta E)^2 - idF\omega \right] b_r(t) = 0.$$

It is seen that these equations describe a one-dimensional harmonic oscillator. Therefore it is not hard to find their solutions.

We introduce the new independent variable instead of time:

$$\tau \equiv t\sqrt{2dF\omega}$$

and also the dimensionless parameter

$$\gamma \equiv \frac{(\Delta E)^2}{dF\omega}.$$

Then we can rewrite Eqs.(5.33) in the simpler form

$$\frac{d^2 b_l(\tau)}{d\tau^2} + \left[\frac{\tau^2}{4} - A_l\right] b_l(\tau) = 0, \qquad A_l = -\frac{\gamma}{8} - \frac{i}{2};$$

$$\frac{d^2 b_r(\tau)}{d\tau^2} + \left[\frac{\tau^2}{4} - A_r\right] b_r(\tau) = 0, \qquad A_r = -\frac{\gamma}{8} + \frac{i}{2}.$$

(5.34)

Let us find the solution of the second of Eqs.(5.34) under the boundary condition

$$\left|b_r(-\infty)\right| = 1. \tag{5.35}$$

We give the physical explanation of this condition below. According to Ref. [4, Sect. 19.7.6] we find the solution in the form of one of the complex *functions of a parabolic cylinder*

$$b_r(t) = \text{const} \cdot E(A_r, \tau). \tag{5.36}$$

According to Sect. 19.21.1 of the same book we find asymptotic representation of this solution:

$$E\left(A_r, \tau \to \pm\infty\right) \approx \sqrt{\frac{2}{\tau}} \exp\left\{ i\left[\frac{\tau^2}{4} - A_r \ln \tau + \Phi + \frac{\pi}{4}\right]\right\}. \tag{5.37}$$

Here the notation

$$\Phi = \frac{1}{2} \arg \Gamma\left(\frac{1}{2} + iA_r\right)$$

is introduced. The term $+i/2$ in A_r produces the factor $\tau^{1/2}$ in Eq.(5.37) which cancels the factor $\tau^{-1/2}$ before the exponent in this expression. Thus, this function of a parabolic cylinder is a constant at infinity, which is in agreement with the condition (5.35).

It follows from Eqs.(5.36) and (5.37) that

$$
\left| \frac{b_r(+\infty)}{b_r(-\infty)} \right|^2 = \left| \frac{\exp\left[-iA_r \ \ln(+\tau)\right]}{\exp\left[-iA_r \ \ln(-\tau)\right]} \right|^2 = \exp\left(-\frac{\pi\gamma}{4}\right). \tag{5.38}
$$

Here we have used the above expressions for A_r from Eqs. (5.34).

Now we analyse the obtained solution. If we consider time as a slowly changing independent variable, then we can approximate the solution of the system (5.32) in the simple form

$$
b_l(t), \ b_r(t) \propto \exp\left[-i\int_0^t E(t')dt'\right]. \tag{5.39}
$$

The quantity $E(t)$ is called the *adiabatic energy*. Substituting Eq.(5.39) into Eq.(5.32) we obtain two values of the adiabatic energy:

$$
E_\pm(t) = \pm\sqrt{\left(\frac{\Delta E}{2}\right)^2 + (dF\omega t)^2}. \tag{5.40}
$$

Of course, the same values can be obtained from the analogous solution of the system (5.30).

Let us assume that as $t \to -\infty$ the lower energy E_- corresponds to the solution b_l while the upper energy E_+ corresponds to the solution b_r. Then at $t = 0$ these levels coincide the unperturbed levels "1"and "2"of the two-level system, respectively. Further as $t \to +\infty$ the lower adiabatic energy E_- corresponds to solution b_r while the upper adiabatic energy E_+ corresponds to solution b_l. Indeed, the solutions of Eqs.(5.32) correspond to the so-called *diabatic energies* $\pm dF\omega t$ when the coupling moment $\Delta E/2$ vanishes.

We can conclude from this consideration that the transition probability between the unperturbed levels "1" and "2" is given by Eq. (5.38):

$$W_{12} = \exp\left(-\frac{\pi(\Delta E)^2}{4dF\omega}\right) \qquad (5.41)$$

This transition is called *the Landau - Zener transition*. It is absent if $F = 0$, as it should be.

We should keep in mind that here we have considered only the vicinity of the time $t = 0$. Analogous transitions within one period of the electromagnetic field take place near the time $t = \pi/\omega$. The lower adiabatic energy curve on the temporal interval between $t < 0$ and $t > \pi/\omega$. corresponds to the solutions $b_r \to b_l \to b_r$ while the upper adiabatic energy curve corresponds to the solutions $b_l \to b_r \to b_l$ According to Eq.(5.41) the probability of remaining on the same unperturbed level is

$$W_{11} = 1 - \exp\left(-\frac{\pi(\Delta E)^2}{4dF\omega}\right) \qquad (5.42)$$

It is equal to 1 at $F = 0$, as it should be.

A transition via two crossing points can be achieved in two ways: $b_r \to b_r \to b_l$ and $b_r \to b_l \to b_l$. In both cases the transition from the lower level to the upper level of the considered two-level system occurs. Each probability is obtained by multiplying Eq. (5.41) by Eq.(5.42). Both probabilities are equal to each other. Thus we finally obtain the so-called *Landau - Zener formula*

$$W_{trans} = 2\exp\left(-\frac{\pi(\Delta E)^2}{4dF\omega}\right)\cdot\left\{1 - \exp\left(-\frac{\pi(\Delta E)^2}{4dF\omega}\right)\right\}. \qquad (5.43)$$

Analogously we can consider the contribution of other periods of the electromagnetic field (see Problem 3).

5.5 THE DIMENSION OF THE SPACE AS A LARGE PARAMETER

Here we consider the case when the dimension of the space (i.e., $N = 3$ for our three-dimensional world) is a large parameter, so that we can expand the solutions of three-dimensional problems over $1/N \ll 1$.

As an example we consider the three-dimensional Schroedinger equation of quantum mechanics for s-states in the central potential $U(r)$. In the general case of N –dimensional space the Laplace operator for spherical symmetric functions is of the form

$$\Delta\psi = \frac{\mathrm{d}^2\psi}{\mathrm{d}r^2} + \frac{N-1}{r}\frac{\mathrm{d}\psi}{\mathrm{d}r}; \qquad r = \sum_{i=1}^{N} x_i^2 . \qquad (5.44)$$

This result can be obtained using the relation $\Delta = \mathrm{div\, grad}$, and also grad $\psi(r) = (\mathrm{d}\psi/\mathrm{d}r)(\mathbf{r}/r)$ and div $\mathbf{r} = \mathrm{N}$. Therefore the Schroedinger equation

$$-\frac{1}{2}\Delta\psi + U(r)\psi = E\psi, \qquad (m = \hbar = 1) \qquad (5.45)$$

is reduced to the one-dimensional wave problem for s-states by the substitution

$$\psi = \frac{\chi(r)}{r^{(N-1)/2}}.$$

After simple derivations we obtain the one-dimensional Schroedinger equation

$$-\frac{1}{2}\chi'' + U_{\mathrm{eff}}(r)\chi = E\chi. \qquad (5.46)$$

Here the effective potential for s - states is of the form

$$U_{\text{eff}}(r) \equiv U(r) + \frac{(N-1)(N-3)}{8r^2}.$$ (5.47)

It differs from the initial potential by the *centrifugal potential* in N-dimensional space. In the limit of large values of $N \gg 1$ we obtain the expansion of (5.47) in powers of $1/N \ll 1$:

$$U_{\text{eff}}(r) = U(r) + \frac{N^2}{8r^2}\left(1 - \frac{4}{N} + \ldots\right)$$ (5.48)

In the case of a super-intense perturbation the potential energy $U(r)$ should be very large. Thus, according to (5.48) it can be represented in the form that corresponds to the principal idea of the $1/N$ approach:

$$U(r) \equiv N^2 v(r),$$ (5.49)

thus, the function $v(r)$ does not contain the parameter of the expansion N.

Now we expand the effective potential $U_{\text{eff}}(r)$ in (5.48) analogously to the general expansion (5.3) near its minimum:

$$U_{\text{eff}}(r) \approx N^2\left\{v(r_0) + (r - r_0)v'(r_0) + \frac{1}{2}v''(r_0)(r - r_0)^2 + \ldots\right\}$$

$$+ \frac{N^2}{8}\left(1 - \frac{4}{N}\right)\left\{\frac{1}{r_0^2} - \frac{2(r - r_0)}{r_0^3} + \frac{3(r - r_0)^2}{r_0^4} + \ldots\right\}.$$ (5.50)

The condition of the minimum of this effective potential energy is

$$U'_{\text{eff}}(r_0) = 0, \qquad \text{i.e.} \qquad v'(r_0) = \frac{1}{4r_0^3}.$$ (5.51)

The energy E in (5.45) is expanded in powers of $1/N$: $E = E^{(0)} + E^{(1)} + \ldots$
Thus we find from (5.50) for the zero approximation of the energy eigenvalue:

$$E^{(0)} = N^2 \left\{ v(r_0) + \frac{1}{8r_0^2} \right\}. \tag{5.52}$$

The first approximation in (5.50) corresponds to a quantum harmonic oscillator approximation with frequency (analogously to (5.3)):

$$\omega = N \sqrt{v''(r_0) + \frac{3}{4r_0^4}}. \tag{5.53}$$

Hence, the first approximation for the energy eigenvalue is the quantum energy of the oscillator, i.e.

$$E^{(1)} = \left(n_r + \frac{1}{2} \right) \omega - \frac{N}{2r_0^2}, \qquad n_r = 0, 1, 2, \ldots \tag{5.54}$$

(the last term on the right side of (5.54) appears from the second part of the expression (5.50)). In quantum mechanics the integer n_r is called the *radial quantum number*.

Finally, the approximate eigenvalues of the differential equation (5.45) are given by the expression

$$E \approx N^2 \left\{ v(r_0) + \frac{1}{8r_0^2} \right\} + N \left\{ \left(n_r + \frac{1}{2} \right) \sqrt{v''(r_0) + \frac{3}{4r_0^4}} - \frac{1}{2r_0^2} \right\} + \ldots \tag{5.55}$$

5.6 AN EXAMPLE OF A LINEAR POTENTIAL

Let us consider a simple example of the application of the results of the previous section: the linear potential energy

$$U(r) = Ar.$$

According to (5.49) we introduce the function

$$v(r) = \frac{A}{N^2} r.$$

According to (5.51) we find the position of the minimum effective potential energy:

$$r_0 = \sqrt[3]{\frac{N^2}{4A}}.$$

The frequency of the harmonic oscillator can be derived from Eq. (5.53) taking into account that $d^2v/dr^2 = 0$:

$$\omega = \sqrt{3} \sqrt[3]{\frac{2A^2}{N}}.$$

The zero approximation for the energy is derived according to Eq. (5.52):

$$E^{(0)} = \frac{3}{2} \left(\frac{AN}{2} \right)^{2/3}.$$

The first approximation for the energy of the ground state $(n_r = 0)$ is derived according to Eq. (5.54):

$$E^{(1)} = \left(\sqrt{3} - 2 \right) \sqrt[3]{\frac{A^2}{4N}}.$$

Thus, using the last two formulae we finally obtain the energy of the ground state in the general form

$$
E = \frac{3}{2}\left(\frac{AN}{2}\right)^{2/3}\left(1 - \frac{0.1786}{N} + \ldots\right) \tag{5.56}
$$

Analogously we can derive the eigenvalues of the energies for the excited states, and also we can find the next iterations of the $1/N$ expansion.

In the case of a three-dimensional problem we substitute $N = 3$ into (5.56). Then we obtain $E^{\text{approx}} \approx 1.848A^{2/3}$. It should be noted that the exact numerical value of the energy of the ground state in the linear three-dimensional potential $U(r) = Ar$ is $E^{\text{exact}} = 1.856A^{2/3}$. The high accuracy of the $1/N$ approximation is seen, though the value of $N = 3$ is not high.

In conclusion we even consider the one-dimensional problem, $N = 1$. The potential energy is of the form $U(x) = A|x|$. According to (5.56) we obtain, substituting $N = 1$: $E^{\text{approx}} \approx 0.776A^{2/3}$. The exact numerical value of this energy is $E^{\text{exact}} = 0.809A^{2/3}$. High accuracy is also seen, though in this case the expansion parameter is $N = 1$! It should be noted that using the second iteration on $1/N$, we will obtain much higher accuracy.

PROBLEMS

Problem 1. Calculate exact and approximate (Eq.(5.16)) energies of the ground state in the potential of the form

$$
U(x) = \left(\frac{1}{x} - x\right)^2
$$

and show that both values differ from one another only by 2.6%.

Problem 2. Calculate exact and approximate (according to Eq.(5.16)) energies of the states with quantum number m in the Morse potential

$$
U(x) = U_0\{\exp(-2x) - 2\exp(-x)\}.
$$

Answer:

$$E(m)_{exact} = -\frac{1}{8}\{2\sqrt{2U_0} - (2m+1)\}^2;$$

$$E(m)_{approx} \approx -U_0 + \sqrt{2U_0}\left(m + \frac{1}{2}\right) - \frac{1}{2}\left(m + \frac{1}{2}\right)^2 + \dots$$

The approximate expression can be obtained from the exact expression by means of the expansion with respect to the small parameter $(2U_0)^{-1/2}$.

Problem 3. Generalize Eq.(5.43) to an arbitrary number of periods of the external monochromatic fields and find the transition rate.

Problem 4. Calculate the oscillating factor in the Landau – Zener transition probability due to the interference between neighbouring crossing points (the so-called *Stueckelberg correction*).

Problem 5. In the limit $\Delta E \to 0$ derive the Landau – Zener transition probability in the framework of the usual perturbation theory. Check that the obtained result coincides with the limit of Eq.(5.43).

Problem 6. Obtain the limit of Eq.(5.43) under the condition of a quasi-stationary perturbation

$$(\Delta E)^2 \gg dF\omega$$

in the form

$$W_{trans} = 2 \exp\left(-\frac{(\Delta E)^2}{4\,dF\omega}\right) \ll 1,$$

Hint: use the WKB approximation.

Problem 7. Using the 1/N expansion (Sect. 5.5), show that in the case of an N-dimensional harmonic oscillator potential

$$U(r) = \frac{1}{2}\omega_0^2 r^2 \, ,$$

the approximate value of the energy coincides with the exact value

$$E(n_r) = \left(2n_r + \frac{N}{2} \right)\omega_0 \, .$$

Here $n_r = 0, 1, 2, \ldots$

Chapter 6

Inverse Problems

6.1 CLASSICAL MECHANICS

This chapter is devoted to the restoration of the coefficients of differential equations knowing their solutions, i.e. so-called *inverse problems*. We consider here various examples in theoretical physics and begin with the analysis of Newton's equations in classical mechanics.

We choose as an example the scattering problem in classical mechanics. Usually we solve the direct problem, i.e. the scattering potential is known and we must derive the scattering amplitude. The inverse problem is formulated in the following manner: we know the differential cross-section of scattering (for example, from experimental data) and we must restore the form of the scattering potential. In addition, some other quantities may be known: scattering phases, scattering amplitude as a function of scattering angle, scattering amplitude as a function of the energy of the scattered particle, and so on.

The differential cross-section of scattering can be expressed via the impact parameter ρ or via the scattering angle χ :

$$d\sigma = 2\pi\rho d\rho = 2\pi \frac{d\rho}{d\chi} d\chi.$$

In classical mechanics the scattering angle and the impact parameter are connected with each other via a single-valued function.

89

The picture of classical scattering is shown in Fig. 5. In the polar system of coordinates the scattering is determined by the distance r from the scattered particle to the scattering centre and by the angle φ. We assume that the scattering potential $U(r)$ depends only on the distance r. In Fig. 5 we show also the scattering angle χ and the angle φ_0 which corresponds to the minimum distance between the particle and the scattering centre. Obviously, $\chi = \pi - 2\varphi_0$.

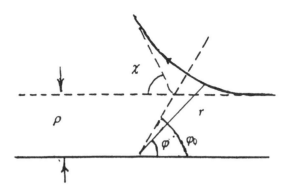

Figure 5. Classical scattering of a particle with the impact parameter ρ and scattering angle χ.

The conservation laws of angular momentum and energy are of the form

$$M = mr^2 \frac{d\varphi}{dt};$$

$$E = \frac{1}{2}m\left(\frac{dr}{dt}\right)^2 + \frac{1}{2}mr^2\left(\frac{d\varphi}{dt}\right)^2 + U(r).$$

(6.1)

Here m is the mass of the scattered particle, M is the angular momentum, E is the total energy, and $U(r)$ is the scattering potential.

The second expression in Eq.(6.1) can be rewritten taking into account the first expression:

$$E = \frac{1}{2} m \left(\frac{dr}{dt} \right)^2 + \frac{M^2}{2mr^2} + U(r).$$

It follows from this expression that the radial velocity of the particle is

$$\frac{dr}{dt} = \sqrt{\frac{2}{m}[E - U(r)] - \frac{M^2}{m^2 r^2}}.$$

Dividing this expression by the first of the expressions (6.1) we find

$$\frac{d\varphi}{dr} = \frac{M}{r^2 \sqrt{2m[E - U(r)] - \frac{M^2}{r^2}}}. \tag{6.2}$$

Instead of the total energy of the particle E and its angular momentum M we introduce the velocity of the particle at infinity, v, and the impact parameter ρ :

$$E = \frac{1}{2} mv^2, \qquad\qquad M = mv\rho.$$

Integrating Eq.(6.2) from infinity up to the minimum distance between the particle and scattering centre we obtain:

$$\varphi_0 = \frac{1}{2} (\pi - \chi) = \int\limits_{r_{min}}^{\infty} \frac{\rho dr}{r^2 \sqrt{1 - \frac{U(r)}{E} - \frac{\rho^2}{r^2}}}. \tag{6.3}$$

This expression is the solution of the direct scattering problem: the scattering potential $U(r)$ is known, and the scattering angle χ is derived as a function of the impact parameter ρ .

Let us now consider the inverse scattering problem. We should derive the scattering potential $U(r)$ knowing the dependence $\chi = \chi(\rho)$, that is, knowing the differential cross-section of scattering. We assume that the scattering potential is of the sufficiently simple form of a monotonically decreasing curve which corresponds to repulsion of the particle from the centre (Fig. 6).

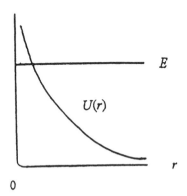

Figure 6. The qualitative form of the scattering potential $U(r)$.

The particle with positive energy E is scattered by this potential, and we should solve Eq. (6.3) with respect to the potential $U(r)$, i.e. the so-called *integral Abel equation*.

In order to solve this problem we introduce other variables instead of U and ρ:

$$s \equiv \frac{1}{r}, \qquad x \equiv \frac{1}{\rho^2}, \qquad w(s) \equiv \sqrt{1 - \frac{U(s)}{E}}.$$

Then Eq.(6.3) can be rewritten in the simpler form

$$\frac{1}{2}(\pi - \chi) = \int_0^{s_0} \frac{ds}{\sqrt{xw^2 - s^2}}. \tag{6.4}$$

Here the upper limit of integration is found from the equation

$$xw^2(s_0) - s_o^2 = 0.$$
(6.5)

Abel was the first to suggest multiplying Eq.(6.4) by the factor

$$\frac{1}{\sqrt{\alpha - x}}$$

and integrating over x in the limits from 0 to α. This procedure can be explained by the fact that integration over s in the right side of Eq.(6.4) produces the simple function:

$$\int_0^\alpha \frac{(\pi - \chi)dx}{2\sqrt{\alpha - x}} = \int_0^{s_0(\alpha)} \int_0^\alpha \frac{dx \cdot ds}{x_o(s)\sqrt{[xw^2(s) - s^2](\alpha - x)}} = \pi \int_0^{s_0(\alpha)} \frac{ds}{w(s)}.$$
(6.6)

The lower limit of integration $x_0(s)$ in Eq.(6.6) is found from the equation

$$x_o w^2(s) - s^2 = 0.$$

The interchange of the integration limits in Eq.(6.6) can be explained using Fig. 7.

Firstly the integration is taken over the variable s in the limits $[0, s_0(x)]$ and secondly over the variable x in the limits $[0, \alpha]$. After interchange of the integration limits according to Fig. 7 we obtain that firstly integration is taken over the variable x in the limits $[x_0(s), \alpha]$ and secondly over the variable s in the limits $[0, s_0(\alpha)]$ where the upper limit of integration is found from the equation

$$\alpha w^2(s_0) - s_o^2 = 0$$

(see Fig. 7 and Eq.(6.5)).

The left side of Eq.(6.6) is the difference of two integrals. The first integral is simply derived and it is equal to $\pi(\alpha)^{1/2}$. Integrating the second integral by parts using the condition $\chi(x = 0) = 0$ (see Fig. 7) we can rewrite Eq.(6.6) in the form

$$\pi\sqrt{\alpha} - \int_0^{\alpha}\sqrt{\alpha - x}\,\frac{d\chi}{dx}\,dx = \pi\int_0^{s_0(\alpha)}\frac{ds}{w(s)}.$$ (6.7)

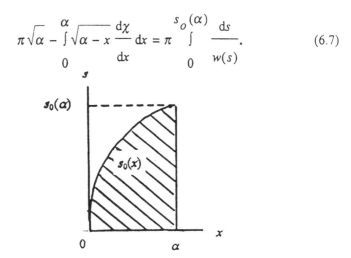

Figure 7. The region of the variables for the double integration in Eq.(6.6).

In order to evaluate the integral on the right side of Eq.(6.7) we differentiate this equation with respect to the variable α :

$$\frac{\pi}{2\sqrt{\alpha}} - \int_0^{\alpha}\frac{1}{2\sqrt{\alpha - x}}\,\frac{d\chi}{dx}\,dx = \frac{\pi}{w(s_0(\alpha))}\,\frac{ds_0}{d\alpha}.$$ (6.8)

Further we change $s_0(\alpha)$ for the sake of simplicity. Then we obtain that

$$\alpha = \left(\frac{s}{w}\right)^2.$$

Substituting this expression into Eq.(6.8) we rewrite it in the form

$$\frac{\pi w}{2s} - \int_0^{(s/w)^2}\frac{\chi'(x)dx}{2\sqrt{(s/w)^2 - x}} = \frac{\pi}{w}\,\frac{ds}{d\alpha}.$$ (6.9)

Let us multiply this equation by the quantity

$$\frac{d\alpha}{ds} = 2\frac{s}{w}\frac{d}{ds}\left(\frac{s}{w}\right).$$

Then we obtain the integral equation

$$\pi\frac{d}{ds}\left(\frac{s}{w}\right) - 2\frac{s}{w}\frac{d}{ds}\left(\frac{s}{w}\right)\int_0^{(s/w)^2}\frac{\chi'(x)dx}{2\sqrt{(s/w)^2 - x}} - \frac{\pi}{w} = 0.$$

The first and third terms in the left side of this equation can be unified; then we rewrite it in the form

$$\pi\frac{d(\ln w)}{ds} = -\frac{d}{ds}\left(\frac{s}{w}\right)\int_0^{(s/w)^2}\frac{\chi'(x)dx}{\sqrt{(s/w)^2 - x}}.$$

Let us introduce the independent variable $u = s/w$ instead of the variable s. Then we can rewrite the last equation in the simpler form

$$\pi\cdot d(\ln w) = -\int_0^{u^2}\frac{\chi'(x)dx}{\sqrt{u^2 - x}}\cdot du. \tag{6.10}$$

Let us integrate this equation over the variable u in the limits $[0, u]$. The double integral on the right side of this equation is derived over the integration region which is shown in Fig. 8.

In the derivation we also take into account that $w(s=0) = 1$ and $\ln w(s=0) = 0$. Rearranging the integration over the variables x and u in the right side of Eq.(6.10), we obtain

$$\pi\cdot\ln w(u) = -\int_0^{u^2}\chi'(x)dx\int_{\sqrt{x}}^u\frac{du}{\sqrt{u^2 - x}}.$$

The new integration limits follow from Fig. 8. After integration over the variable u we find

$$\pi \cdot \ln w(u) = - \int_0^{u^2} \chi'(x) \cdot \mathrm{arccosh}\left(\frac{u}{\sqrt{x}}\right) dx.$$

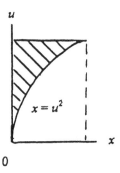

Figure 8. The integration region for the integral on the right side of Eq. (6.10).

Returning from the variable x to the variable ρ, we can rewrite this equation in the form:

$$\pi \ln w(\rho_0) = \int_{\rho_0}^{\infty} \frac{d\chi}{d\rho} \,\mathrm{arccosh}\left(\frac{\rho}{wr}\right) d\rho. \qquad (6.11)$$

Here the notation

$$\rho_0 = wr = r\sqrt{1 - \frac{U(r)}{E}}.$$

is introduced (according to Eq.(6.5)). Integrating the right side of Eq.(6.11) by parts we obtain

$$\pi \ln w = - \int\limits_{wr}^{\infty} \frac{\chi(\rho)d\rho}{\sqrt{\rho^2 - (wr)^2}}.$$

Thus, finally we obtain the equation

$$w(r) = \exp\left\{ -\frac{1}{\pi} \int\limits_{wr}^{\infty} \frac{\chi(\rho)d\rho}{\sqrt{\rho^2 - (wr)^2}} \right\}. \qquad (6.12)$$

This equation determines the function $w(r)$ and, hence, the potential $U(r)$. It should be noted that this function is contained also in the right side of Eq.(6.12). Thus, we complete the solution of the inverse problem since Eq.(6.12) allows us in principle to derive the potential $U(r)$ if the dependence $\chi(\rho)$ of the scattering angle·on the impact parameter ρ is known.

Let us remember that our calculations are valid if this potential is repulsive and decreases monotonically with the distance r.

6.2 THE EXAMPLE OF THE RUTHERFORD FORMULA

Let us check the general equation (6.12) on the example of Coulomb scattering. For the sake of simplicity we restrict ourselves to the case of small scattering angles. Firstly let us find the dependence of the scattering angle on the impact parameter for the scattering of a particle with the mass m by the potential $U(r) = Z/r$. The scattering trajectory is shown in Fig. 9.

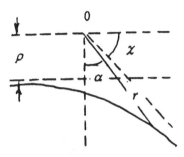

Figure 9. Coulomb scattering at a small angle.

Let us calculate the momentum of the particle produced due to the scattering in the direction which is normal to the initial direction of this particle:

$$
p_\perp = \int_{-\infty}^{\infty} F_\perp \, dt = \int_{-\infty}^{\infty} \frac{Z}{r^2} \cos\alpha \cdot dt
$$

$$
= \int_{-\infty}^{\infty} \frac{Z\rho \, dt}{(\rho^2 + v^2 t^2)^{3/2}} = \frac{2Z}{\rho v} \int_0^{\infty} \frac{dx}{(1 + x^2)^{3/2}} = \frac{2Z}{\rho v}.
$$

On the other hand, the normal projection of the momentum can be expressed via its momentum p and the small scattering angle χ (see Fig. 9): $p_\perp = p\chi$, $\chi \ll 1$. Then we can determine the dependence of the scattering angle on the impact parameter in the case of Coulomb scattering at small angles:

$$
\chi = \frac{2Z}{pv\rho} = \frac{Z}{\rho E}, \tag{6.13}
$$

where $E = mv^2/2 = pv/2$ is the energy of the particle at infinity (this is the *Rutherford formula*).

Substituting Eq.(6.13) in the general Eq.(6.12) we can write the right side in the form

$$
\int_{wr}^{\infty} \frac{Z \, d\rho}{\rho E \sqrt{\rho^2 - (wr)^2}} = \frac{\pi Z}{2Ewr}.
$$

Hence, it follows from Eq.(6.12) that

$$
w = \exp\left(-\frac{Z}{2Ewr}\right) \approx 1 - \frac{Z}{2Ewr} \approx 1 - \frac{Z}{2Er}
$$

since in the case of large values of the impact parameter the potential energy of the Coulomb interaction is small compared to the kinetic energy of the particle. According to the definition we have

$$w = \sqrt{1 - \frac{U(r)}{E}} \approx 1 - \frac{U(r)}{2E}.$$

Comparing two last expressions we finally obtain that

$$U(r) = \frac{Z}{r},$$

as it should be.

Thus we derive an explicit expression for the Coulomb potential from the dependence of the scattering angle on the impact parameter (in the case of small scattering angles), i.e. we solve the inverse classical problem for this simple physical example.

6.3 THE BORN APPROXIMATION

In this section we consider a case which is opposite to the case which was investigated in the previous section. Now we suggest that a particle with large   energy is scattered by some central potential of the type shown in Fig. 6.

It is well known that under these conditions the *Born approximation* of quantum mechanics is applicable. According to this approximation the scattering amplitude f is determined by the scattering potential $U(r)$ in first order perturbation theory:

$$f(q) = -\frac{1}{2\pi} \int U(r) \exp(-i\mathbf{q}\mathbf{r})d\mathbf{r}. \qquad (6.14)$$

Here the transferred momentum q is determined by the relation

$$q = 2p \sin(\chi / 2),$$

where p is the momentum of the particle and χ is the scattering angle. For the sake of simplicity we use the system of units where the mass of a particle m and the Planck constant are equal to unit.

We should now express the scattering potential via the scattering amplitude. This is achieved by the inverse Fourier transformation. Thus, we multiply Eq.(6.14) by the exponent $\exp(i\mathbf{q}\mathbf{r}')$ and integrate over $\mathbf{q}$. Then we obtain

$$- \frac{1}{2\pi} \int U(r) \exp(i\mathbf{q}\mathbf{r}' - i\mathbf{q}\mathbf{r}) d\mathbf{q} d\mathbf{r}' = - \frac{1}{2\pi} (2\pi)^3 U(r')$$

$$= \int f(q) \exp(i\mathbf{q}\mathbf{r}') d\mathbf{q} = \frac{4\pi}{r'} \int_0^\infty f(q) \sin(qr') dq.$$

The final expression for the scattering potential is of the form

$$U(r) = - \frac{1}{2\pi} \int_0^\infty f(q) \sin(qr) q \cdot dq. \qquad (6.15)$$

It is seen that in order to derive the scattering potential we must know the dependence of the differential cross-section of scattering on the energy at some fixed value of the scattering angle χ. This dependence must be known in a sufficiently wide interval of energies in order to obtain the correct value of the integral in Eq.(6.15). The scattering potential changes over the typical length R which is of the order of $1/Q$, where Q is the typical transferred momentum for the experimentally measured scattering amplitude. If the scattering angle is $\chi \sim 1$ then the value of $Q \sim p = (2E)^{1/2}$, where E is the energy of the particle. However if the scattering angle is $\chi \ll 1$ (scattering on small angles) then the value of Q is much less than the above estimate. Hence, the value of R is much larger than in the previous case. Thus scattering over different angles allows us to derive the scattering potential for different distances.

6.4 PHASE THEORY OF SCATTERING

In this section we consider the method of restoration of the scattering potential knowing the scattering phase. According to the phase theory of scattering in quantum mechanics [2] the scattering amplitude f is expressed via the partial scattering phase δ_l (l is the orbital quantum number) :

$$f(\chi) = \sum_{l=0}^\infty \frac{\exp(2i\delta_l) - 1}{2ip} P_l(\cos \chi). \qquad (6.16)$$

Here p is the momentum of the particle and χ is the scattering angle. We assume that the scattering phases are small compared to 1 and that the zero scattering phase $\delta_0 \ll 1$ is known from experimental data. These conditions correspond to the applicability of perturbation theory, i.e. of the Born approximation. Hence, the Born formula (6.14) is correct.

Equating expressions (6.14) and (6.16) to each other and integrating the obtained equation over the solid scattering angle Ω in order to exclude all phases except the zero phase from Eq.(6.16) we find

$$4\pi \frac{\delta_0}{p} = -\frac{1}{2\pi} \int d\Omega \int U(r) \exp(-i\mathbf{q}\mathbf{r}) d\mathbf{r}.$$

Since the transferred momentum q is connected with the scattering angle χ by the relation

$$q^2 = 2p^2 \left(1 - \cos\chi\right), \qquad\qquad 2q\,dq = 2p^2 \sin\chi\,d\chi,$$

the integration over the scattering angle can be changed by integration over the transferred momentum at the fixed energy of the particle:

$$\delta_0 = -\frac{1}{p} \int_0^\infty dr \int_0^{2p} \sin(qr)\,dq \cdot rU(r).$$

Integrating this expression over q we finally obtain

$$\delta_0 = -\frac{2}{p} \int_0^\infty U(r) \sin^2(pr)\,dr. \qquad\qquad (6.17)$$

In order to reverse this dependence we multiply it by the momentum of the particle p and differentiate with respect to this momentum:

$$\frac{d}{dp}(p\delta_0) = -2 \int_0^\infty rU(r) \sin(2pr)\,dr = -i \int_0^\infty rU(r) \exp(-2ipr)\,dr.$$

In the right side of this expression we have the inverse Fourier component. The inverse Fourier transformation gives the result:

$$rU(r) = \frac{i}{2\pi} \int\limits_{-\infty}^{\infty} \frac{d}{dp}(p\delta_0)\exp(2ipr)\cdot d(2p).$$

According to Eq.(6.17) the zero scattering phase is an odd function of the momentum. Hence, we can replace the exponent in this expression by its odd part, i.e. by $\sin(2pr)$. Finally we obtain the simple expression for the scattering potential:

$$U(r) = -\frac{2}{\pi r} \int\limits_{0}^{\infty} \frac{d}{dp}(p\delta_0)\sin(2pr)dp. \qquad (6.18)$$

Thus, we have restored the scattering potential via the zero scattering phase. Let us remember that this result is valid when the scattering phase is small compared to 1. In the case when the scattering phase is not small, this approach allows us to carry out the iteration procedure where Eq.(6.18) is the first term of iterations for the scattering potential.

6.5 INVERSION OF THE PERTURBATION THEORY FOR EIGENVALUE PROBLEMS

In the framework of first order stationary perturbation theory we can simply derive shifts of eigenvalues of the differential equation: this is the diagonal matrix element of the perturbative potential taken on the unperturbed functions for the considered state. The inverse problem is the derivation of the weak perturbative potential knowing all small shifts of eigenvalues.

For example, let us consider the simplest problem in quantum mechanics – the eigenvalues of the energy and the eigenfunctions in the infinite one-dimensional rectangular well. The unperturbed potential is of the form

$$U(x) = \begin{cases} 0, & 0 < x < a \\ \infty, & x < 0, x > a. \end{cases}$$

The eigenfunctions and the energies of the one-dimensional Schroedinger equation

$$-\frac{\hbar^2}{2m}\frac{d^2\psi_n}{dx^2} + U(x)\psi_n = E_n\psi_n$$

are of the well-known form

$$\psi_n(x) = \sqrt{\frac{2}{a}}\sin\left(\frac{\pi nx}{a}\right); \qquad E_n = \frac{\pi^2 n^2}{2ma^2}; \qquad n = 1,2,3...$$

The shifts of the eigenvalues produced by a weak perturbative potential $\delta U(x)$ are derived in the first-order perturbation theory as the diagonal matrix element of this perturbative potential:

$$\delta E_n = \int_0^a \delta U(x)\psi_n^2(x)dx = \frac{1}{a}\int_0^a \delta U(x)\left(1 - \cos\frac{2\pi nx}{a}\right)dx.$$

If the quantities δE_n are known, then we can derive the perturbative potential starting from this equation

$$\delta U(x) = -2\sum_{n=1}^{\infty} \delta E_n \cos\frac{2\pi nx}{a}. \qquad (6.19)$$

In particular, if, for example, only the energy of the ground state is weakly shifted, then the perturbative potential is

$$\delta U(x) = -2\delta E_1 \cos\frac{2\pi x}{a}.$$

These results can be generalized to the case of an arbitrary potential considered in the framework of the WKB approximation (see Chapter 2). The WKB wave functions are of the form

$$\psi_n(x) \propto \cos\left(\int_a^x p_n(x')dx' - \frac{\pi}{4}\right)$$

Here the WKB momentum of a particle is determined as

$$p_n(x) = \sqrt{2m(E_n - U(x))}, \qquad \hbar = 1.$$

The quantities E_n and $U(x)$ are the energy eigenvalue and the unperturbed potential, respectively.

Instead of (6.19) we obtain now

$$\delta U(x) = -2\Sigma_n \delta E_n \sin\left(2\int_a^x p_n(x')dx'\right)$$

Of course, we have in this case only approximate orthogonality of eigenfunctions with different energies, unlike the previous case of the rectangular well.

PROBLEMS

Problem 1. Let us assume that the zero scattering phase is known in the approximation of the effective radius, i.e.

$$p \cot \delta_0 = -\frac{1}{a} + \frac{1}{2} r_0 p^2.$$

Here a is the so-called *scattering length* , and r_0 is the so-called *effective radius of interaction.* The value of a is supposed to be negative. Obtain the scattering potential from Eq.(6.18).

Answer:

$$U(r) = -\frac{8}{r_0 R} \exp\left(-\frac{2r}{R}\right)$$

where the radius of the potential R is given by the expression

$$R = \sqrt{-\frac{r_0 a}{2}}.$$

Problem 2. Show that if the value of the zero scattering phase is not too small then instead of Eq.(6.18) the first approximation for the scattering potential is of the form

$$U(r) = -\frac{2}{\pi r} \int_0^\infty \frac{d}{dp}(p \tan \delta_0) \sin(2pr) dp.$$

Problem 3. Show that there is large number of so-called *equivalent potentials* which produce the same phases of s-scattering and the same energies of discrete s-states.

Problem 4. Obtain the Coulomb potential of classical scattering from Eq.(6.12) using the Rutherford formula for the differential cross-section of repulsive scattering at arbitrary angles.

Problem 5. Generalize result (6.15) of Sect. 6.3 to the case when both the first and the second terms of the Born approximation should be taken into account.

Problem 6. Let us assume that classical scattering of particles at a centre is isotropic. Using Eq.(6.12) show that the scattering potential represents a hard sphere.

Chapter 7

The Self-Consistent Approximation

7.1 BASIC EQUATIONS OF THE SELF-CONSISTENT APPROXIMATION

The self-consistent approximation is widely used in quantum theory, atomic and solid-state physics, etc. to solve the equations of motion in many-body systems. As a rule, the exact solution of a many-body problem, especially if the particles involved strongly interact with each other, cannot be obtained. Perturbation theory can only be used if the interaction between the particles is weak. If not, the self-consistent approximation often yields good results. This approximation assumes that the action of all particles of the system upon a particle can be replaced by the action of a single-particle potential averaged over the motion of the other particles. Thus, a many-body problem (which is linear with respect to the wave function) is reduced approximately to a single-particle (although often nonlinear) problem. While in the previous chapters we considered linear differential equations, this chapter and the next two chapters will be devoted to the approximate solution of nonlinear differential equations.

If a many-body Schroedinger equation is considered, then a many-body wave function is represented approximately as the product of single-particle wave functions; the potential which the other particles create at the location of the selected particle is averaged over the squared modulus of the wave function of the remaining particles.

In accordance with the previous discussion, the total wave function of the steady state in the self-consistent approximation (the so-called *Hartree approximation*) is

$$\Psi(\mathbf{r}_1, \mathbf{r}_2, ... \mathbf{r}_N) = \psi_1(\mathbf{r}_1) \cdot \psi_2(\mathbf{r}_2) \cdot ... \psi_N(\mathbf{r}_N). \qquad (7.1)$$

The subscripts 1,2,...,N in the single-particle wave functions $\psi_i(\mathbf{r}_i)$ denote the quantum numbers of the single-particle states of the particles.

The stationary Schroedinger equation for a single-particle wave function $\psi_i(\mathbf{r}_i)$ is of the form

$$\left[-\frac{\hbar^2}{2m} \Delta_i + V(\mathbf{r}_i) \right] \psi_i(\mathbf{r}_i) = E_i \psi_i(\mathbf{r}_i). \qquad (7.2)$$

Here Δ_i is the Laplace operator, m is the mass of the ith particle, E_i is the single-particle energy, and the single-particle self-consistent potential $V(\mathbf{r}_i)$ is related to the interaction potential $U(\mathbf{r}_i - \mathbf{r}_j)$ by the relation

$$V(\mathbf{r}_i) = \sum_{j \neq i} \int U(\mathbf{r}_i - \mathbf{r}_j) \left| \psi_j(\mathbf{r}_j) \right|^2 d\vec{r}_j + V_0(\mathbf{r}_i). \qquad (7.3)$$

Here $V_0(\mathbf{r}_i)$ is the potential of the external field, if any, acting upon the particle.

Equation (7.2) can be obtained not only intuitively, but also from *a variational principle* which assumes that the total energy of a many-body problem (the averaged value of the Hamiltonian) is minimized using a trial variational function like in Eq.(7.1). In this case we obtain Eq.(7.2) for single-particle wave functions.

Note that the total energy in the self-consistent approximation is not equal to the sum of the energies of separate particles, because the energy of interaction between any two particles has been counted twice in each single-particle energy. Therefore, the mean energy of the interaction should be subtracted from the total energy, and this must be done for every pair of particles.

The self-consistent approximation, however, has an essential shortcoming. When we resort to an approximation method, we can often estimate the accuracy of the method; for example, the accuracy of the first-order approximation in perturbation theory is determined by the smallness of the second-order term. A similar procedure is applied for the WKB approximation and for other approximate methods. By contrast, the accuracy of the self-consistent approximation cannot be estimated without knowing the exact solution, which is usually impossible to obtain.

Commonly, the accuracy of the self-consistent approximation is estimated   from general assumptions. Firstly, the accuracy of the approximation improves as the number of particles in the system increases. Secondly, our intuition suggests that the approximation is more adequate for systems with a *long-range* interaction between particles (for example, the Coulomb interaction). For this reason, the self-consistent approximation is widely used to investigate atomic structure.

On the other hand, the self-consistent approximation is not optimal for systems with a short-range interaction, when a particle only interacts with one other particle at any moment of time (or does not interact at all). It is interesting to note in this connection that the *shell model* of the nucleus, in which the average self-consistent field of nucleons acts upon a nucleon, adequately describes the structure of low excited states of atomic nuclei, even though the nucleon - nucleon interaction is short range. This can be explained by *Pauli's exclusion principle*. The principle forbids many possible interactions between nucleons because the states which nucleons could transfer to are occupied. The free path of a nucleon increases, which corresponds to an increase in the effective range of nuclear forces, and, therefore, the self-consistent approximation can be applied. Note that the exclusion principle is effective only for low-excited states. Therefore, the shell model can only describe low excited states of middle and heavy atomic nuclei.

The second example is a dilute, magnetically trapped, alkali atom Bose gas with repulsive atomic pair interactions. The phenomena are well described by the mean-field theory of the Bose - Einstein condensate, in which all the gas in the system resides in the condensed state. The condensate wave function is determined by solving the self-consistent equation (the so-called *Gross - Pitaevskii equation*):

$$\left[-\frac{\hbar^2}{2M} \Delta + V_{\text{trap}}(\mathbf{r}) + NU|\psi(\mathbf{r})|^2 \right] \psi(\mathbf{r}) = E\psi(\mathbf{r}). \qquad (7.4)$$

Here the trapping potential (for cylindrically symmetric systems of current interest) is given by

$$V_{\text{trap}}(\mathbf{r}) = \frac{1}{2} M \left(\omega_\rho^2 \rho^2 + \omega_z^2 z^2 \right) \qquad (7.5)$$

with ω_ρ, ω_z, the radial and axial angular frequencies of the trap; the quantity

$$U = \frac{4\pi\hbar^2 a}{M}$$

expresses the binary interaction between atoms in the Born approximation (see Section 6.3), with scattering length $a > 0$, and the energy E, interpreted as the work required to add one more atom to the condensate, is treated as an eigenvalue. N is the concentration of atoms.

The simplest way to solve the nonlinear equation (7.5) is the *variational approximation*. Let us consider, for example, the plane $z = 0$ and choose, for the sake of simplicity, $M = h = 1$. Then Eq. (7.4) takes the approximate form of a nonlinear ordinary differential equation depending on the radial coordinate ρ only:

$$\hat{H}\psi = \left\{ -\frac{1}{2}\frac{d^2}{d\rho^2} - \frac{1}{2\rho}\frac{d}{d\rho} + \frac{1}{2}\omega^2\rho + NU|\psi|^2 \right\}\psi = E\psi. \quad (7.6)$$

We seek the solution of this nonlinear equation in the form of a trial function

$$\psi(\rho) = A \exp\left(-\frac{\rho}{R}\right), \quad (7.7)$$

where R is the trial parameter, and A is the normalized factor. This factor is derived from the normalized condition

$$\int_0^\infty |\psi|^2 2\pi\rho d\rho = 1, \quad (7.8)$$

so that after simple derivation we find $A = (2/\pi R^2)^{1/2}$. The energy of the considered condensate state is found as

$$E = \frac{\langle \psi|\hat{H}|\psi \rangle}{\langle \psi|\psi \rangle} = \frac{\int_0^\infty \psi\hat{H}\psi 2\pi\rho d\rho}{\int_0^\infty |\psi|^2 2\pi\rho d\rho}. \quad (7.9)$$

Substituting (7.7) into (7.9) and calculating the simple integral, we obtain

$$E = E(R) = \frac{1}{4\pi}[3\omega_\rho^2 R^2 / 2 + (1 + NU / \pi) / R^2]. \qquad (7.10)$$

According to the variational method we should derive the minimum value of this expression. Differentiating it with respect to the parameter R, we find the equation

$$3\omega_\rho^2 R = \frac{3}{R^3}\left(1 + \frac{NU}{\pi}\right) \qquad (7.11)$$

Thus, finally we derive the variational parameter

$$R = \left(\frac{1 + 4Na}{\omega_\rho^2}\right)^{1/4}. \qquad (7.12)$$

Also, substituting (7.12) into (7.10), we calculate the approximation expression for the eigenvalue of the energy

$$E = E_{min}(R) = \frac{5}{8\pi}\sqrt{1 + 4Na} \cdot \hbar\omega_\rho. \qquad (7.13)$$

Here N is the number of atoms in the unit of length z in the plane $z = 0$.

Analogously we can also solve the two-dimensional problem taking into account the distribution of the wave function on the variable z . In practice, the variational method is a very good approximation for eigenvalues of the differential equation and for the corresponding eigenfunctions.

Unfortunately, in this case we do not have any analytical solutions. Therefore we cannot compare the variational solution with the exact result. But numerical derivations show that the trial function (7.7) approximates well the exact numerically derived wave function. Also the energy (7.13) approximates well the corresponding numerical result.

7.2 AN EXAMPLE OF THE EXACT SOLUTION

Let us consider a simple quantum-mechanical problem, for which both exact and self-consistent solutions can be found and have simple analytical forms. Then we can estimate the reliability of this approximation, since there are no small parameters, and only numerical agreement of the exact and self-consistent solutions need be considered.

Let us find the ground-state energy and the corresponding wave function of a three-dimensional system consisting of two identical particles bound to a fixed center and to each other by elastic forces.

Without loss of generality, we choose a system of units in which the mass of the particle, Planck's constant, and the elasticity coefficient for the attraction to the center are unity. Let us denote the distances from the first particle to the center and from the second particle to the center as r_1 and r_2, respectively. The Hamiltonian of the system is

$$\hat{H} = -\frac{1}{2}(\Delta_1 + \Delta_2) + \frac{1}{2}(r_1^2 + r_2^2) + \frac{1}{2}k(\mathbf{r_1} - \mathbf{r_2})^2. \qquad (7.14)$$

Here k is the elasticity coefficient for the elastic bond between the particles. It may be both positive and negative. The system may simulate, for example, the helium atom in which Coulomb forces are replaced by elastic forces. In this case, the coefficient k is obviously negative.

The stationary Schroedinger equation is

$$\hat{H}\Psi(\mathbf{r_1}, \mathbf{r_2}) = E\Psi(\mathbf{r_1}, \mathbf{r_2}). \qquad (7.15)$$

We need to find the ground-state energy E and the wave function Ψ by solving Eq.(7.15). In order to separate the variables, we make the substitution

$$\mathbf{r} = \frac{1}{\sqrt{2}}(\mathbf{r_1} - \mathbf{r_2}), \qquad \mathbf{R} = \frac{1}{\sqrt{2}}(\mathbf{r_1} + \mathbf{r_2}). \qquad (7.16)$$

Here $\mathbf{r}$ is the relative coordinate, and $\mathbf{R}$ is the coordinate of the center of mass. Thus, the Hamiltonian (7.14) takes the form

$$\hat{H} = -\frac{1}{2}(\Delta_r + \Delta_R) + \frac{1}{2}(r^2 + R^2) + kr^2. \qquad (7.17)$$

Let us seek the solution of Eq.(7.17) in the form

$$\Psi = \varphi(r)\Phi(R). \tag{7.18}$$

Standard normalized solutions for the ground state of the three-dimensional harmonic oscillator are of the form

$$\Phi(R) = \pi^{-3/4}\exp\left(-\frac{1}{2}R^2\right)$$

$$\varphi(r) = \pi^{-3/4}(2k+1)^{3/8}\exp\left(-\frac{1}{2}(2k+1)^{1/2}r^2\right) \tag{7.19}$$

The energy of the ground state is

$$E = \frac{3}{2}[1 + \sqrt{2k+1}]. \tag{7.20}$$

When the coupling between the particles is zero ($k = 0$), it follows from Eq.(7.20) that $E = 3$ as would be expected, i.e. the total energy is equal to the sum of the energies of the two particles in the ground state of the three-dimensional oscillator.

If the coupling is weak ($k \ll 1$), it is easy to derive from Eq.(7.20) the first order term of perturbation theory

$$E \approx 3 + \frac{3}{2}k. \tag{7.21}$$

Note that the exact solution (7.18) cannot be represented as the product of functions of which one depends on r_1 and the other on r_2. The factorization is only possible when $k = 0$.

Solution (7.20) shows that the energy is real for $k > -1/2$. If $k < -1/2$, the energy is complex. This means that the state considered is no longer bound, because repulsion between the particles exceeds their attraction to the center.

7.3 SELF-CONSISTENT SOLUTION

In this section the same problem as above will be solved in an approximation in which a self-consistent field acts upon a particle. According to Eq.(7.1) the complete wave function is represented as the product of two single-particle wave functions, namely

$$\Psi(r_1, r_2) = \psi(r_1)\psi(r_2). \tag{7.22}$$

The single-particle wave function corresponds to the single-particle ground state, therefore the subscripts on the wave functions in Eq.(7.22) are omitted.

The average potential acting upon the first particle is the sum of the potential of its attraction to the center and the potential of its coupling with the second particle, this potential being averaged over the motion of the second particle (see Eq.(7.3)):

$$V(r_1) = \frac{1}{2} r_1^2 + \frac{k}{2} \int (r_1 - r_2)^2 |\psi(r_2)|^2 dr_2. \tag{7.23}$$

Squaring the binomial, we obtain three integrals. The first is easily calculated by using the normalization condition for the wave function $\psi(r_2)$. The second is zero because the integrand is odd. The third integral is the average of the squared coordinate of the second particle $<r_2^2>$. Thus, we find

$$V(r_1) = \frac{k+1}{2} r_1^2 + \frac{k}{2} < r_2^2 > . \tag{7.24}$$

This is the potential of the three-dimensional harmonic oscillator, so that the single-particle wave function of the ground state is of the form:

$$\psi(r_1) = \pi^{-3/4} (k+1)^{3/8} \exp\left[-\frac{1}{2} \sqrt{k+1} \; r_1^2 \right]. \tag{7.25}$$

The single-particle wave function of the second particle is of an analogous form. Thus, it follows from Eqs.(7.22) and (7.25) that

$$\Psi_{approx}(r_1, r_2) = \pi^{-3/2}(k+1)^{3/4} \exp\left[-\frac{1}{2}\sqrt{k+1}(r_1^2 + r_2^2)\right]$$

$$= \pi^{-3/2}(k+1)^{3/4} \exp\left[-\frac{1}{2}\sqrt{k+1}(r^2 + R^2)\right].$$

$$(7.26)$$

The energy of the ground state in the self-consistent approximation is

$$E_{approx} = \left\langle \Psi_{approx} \left| \hat{H} \right| \Psi_{approx} \right\rangle, \qquad (7.27)$$

where the approximate wave function is determined by Eq.(7.26). In order to calculate this energy we rewrite the Hamiltonian (7.14) in the form:

$$\hat{H} = \left\{-\frac{1}{2}\Delta_1 + \frac{1}{2}(k+1)r_1^2\right\} + \left\{-\frac{1}{2}\Delta_2 + \frac{1}{2}(k+1)r_2^2\right\} - k(r_1 \cdot r_2).$$

$$(7.28)$$

The contribution of the third term in the right side of Eq.(7.28) in Eq.(7.27) is equal to zero. The contribution of the first term is equal to the energy of the ground state of the three-dimensional harmonic oscillator with frequency $(k+1)^{1/2}$, i.e. to $3(k+1)^{1/2}/2$. Thus, substituting Eqs.(7.26) and (7.28) into Eq.(7.27) we obtain

$$E_{approx} = 3\sqrt{k+1}. \qquad (7.29)$$

7.4 COMPARING THE SELF-CONSISTENT APPROXIMATION WITH THE EXACT SOLUTION

Now we can compare the results of the exact calculation and the self-consistent approximation for the energy of the ground state, i.e. Eqs. (7.20) and (7.29). They coincide with each other at $k = 0$ as they should do. If the coupling is weak, i.e. $k \ll 1$ then it is seen from Eqs.(7.21) and (7.29) that

$$E = E_{approx} \approx 3 + \frac{3}{2}k. \qquad (7.30)$$

Fig. 10 shows the exact and approximate energies for arbitrary values of k.

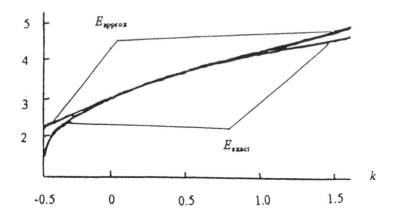

Figure 10. Exact (E_{exact}) and self-consistent (E_{approx}) energies as functions of the coupling parameter k.

It should be noted that even when $k = 1$ the value of the exact energy is about 96.5% of the self-consistent energy! Besides, it is evident from Fig.10 that the energy found from the self-consistent approximation always exceeds the exact energy. This property is common to all variational methods. The conclusion can be drawn that the self-consistent approximation is applicable when the interaction potential between particles is on the order of the magnitude of their kinetic energies, i.e. no small or large parameters are present in the physical problem.

We can now compare the quality of the self-consistent approximation by the derivation of the overlap integral between the exact wave function (7.18) and the approximate wave function (7.26); the square of this integral is

$$I = \left| \left\langle \Psi | \Psi_{approx} \right\rangle \right|^2 = \frac{64(k+1)^{3/2}(2k+1)^{3/4}}{[1 + \sqrt{k+1}]^3 [\sqrt{k+1} + \sqrt{2k+1}]^3}. \quad (7.31)$$

The dependence of the overlap integral on the coupling parameter k is shown in Fig. 11. Of course, the overlap integral is equal to 1.00 for $k = 0$. It should be noted that for $k = 1$ the overlap integral is equal 0.95, i.e. the self-consistent wave function reproduces well the exact wave function. However the overlap decreases for negative values of this parameter, and it is equal to zero for $k = -1/2$, since the exact bound state disappears, and the particles are moving to infinity in opposite directions due to strong repulsion between them.

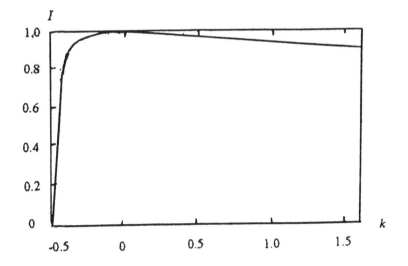

Figure 11. The dependence of the overlap integral I between the exact and self-consistent wave functions on the coupling parameter k.

We have seen that the self-consistent solution has no peculiarities for $k = -1/2$, in contrast to the exact solution. The conclusion can be drawn that the self-consistent approximation describes quite well the domain of parameter values of the order of unity, except for the range of the critical negative coupling constant $k = -1/2$.

7.5 THE SOMMERFELD APPROACH FOR NONLINEAR EQUATIONS

The equations of the self-consistent approximation in quantum mechanics are simplified if a *large number* of particles is involved in the problem. Then the quantum numbers of the single-particle states of these particles are high, so that we can use the WKB approximation (see Chapter 2). Instead of the general case, we consider the example of the derivation of the electrostatic potential in a neutral atom with a large number of electrons. The solution of the nonlinear differential equation will be found by the so-called *Sommerfeld approach*.

The electrostatic Poisson equation for the electrostatic potential φ is

$$\Delta\varphi(r) = 4\pi en(r), \tag{7.32}$$

where n is the electron concentration, and $-e$ is the electron charge. In the WKB approximation these electrons are inside a sphere in momentum space with a radius $p_F(r)$ (this is the so-called Fermi momentum). Each of the electrons populates the volume

$$(2\pi\hbar)^3.$$

Thus, according to the Pauli principle

$$n(r)(2\pi\hbar)^3 = 2\frac{4\pi}{3}p_F^3(r). \tag{7.33}$$

The additional factor of 2 is due to two values of the electron spin in each volume cell.

The kinetic energy of each electron can be connected with its electrostatic energy. The total electron's energy is a constant. The value of this constant is zero for a neutral atom, since at $r\to\infty$ both the kinetic and potential energy of the electron should disappear. Thus,

$$\frac{p_F^2(r)}{2m} - e\varphi(r) = 0. \tag{7.34}$$

Eqs. (7.32-34) lead to the so-called *Thomas–Fermi equation* for the electrostatic potential. Further we use the atomic system of units

$$m = e = \hbar = 1,$$

so this second-order nonlinear differential equation describing the self-consistent distribution of electrons in the neutral atom is of the form

$$\Delta\varphi = \frac{8\sqrt{2}}{3\pi}\varphi^{3/2}. \tag{7.35}$$

The boundary conditions are

$$\varphi(r)\xrightarrow[r\to 0]{}\frac{Z}{r}; \qquad \varphi(r)\xrightarrow[r\to\infty]{}0. \tag{7.36}$$

Here Z is the charge of the atomic nucleus.

Let us introduce the new independent and dependent variables:

$$r = \rho \cdot bZ^{-1/3}; \qquad b = \frac{1}{2}\left(\frac{3\pi}{4}\right)^{2/3}; \qquad \varphi(r) = \frac{Z^{4/3}}{b\rho}\chi(\rho).$$

(7.37)

Then substituting (7.37) into (7.35), we find the universal nonlinear differential equation

$$\frac{d^2\chi}{d\rho^2} = \sqrt{\frac{\chi^3}{\rho}}.$$

(7.38)

The universal boundary conditions are $\chi(0) = 1$; $\chi(\infty) = 0$.

The method of constructing an approximate solution of these equations is by looking at the behavior of the equation near its singular points. *Once we know the solution of a differential equation near its singularities, we can construct an interpolation formula for the solution throughout the whole region of interest (A. Sommerfeld).*

Let us find the approximate solution of (7.38) for $\rho \to \infty$. We shall look for it in the form $\chi(\rho) = A/\rho^\alpha$. Substituting this form of solution in (7.38), we get

$$\chi(\rho) \xrightarrow[\rho \to \infty]{} \frac{144}{\rho^3}.$$

(7.39)

We next find the next-order correction to this solution as $\rho \to \infty$, that is, we put

$$\chi(\rho) \xrightarrow[\rho \to \infty]{} \frac{144}{\rho^3} + \xi(\rho)$$

(7.40)

and solve for ξ. Substituting (7.40) in (7.38), we find the approximation linear in ξ,

$$\frac{d^2\xi}{d\rho^2} = \frac{18}{\rho^2}\xi. \tag{7.41}$$

It is clear from (7.41) that ξ must be some power function of ρ, that is,

$$\xi = B / \rho^k.$$

From (7.41) we get for k the quadratic equation $k(k + 1) = 18$ whence

$$k = \frac{1}{2}(-1 \pm \sqrt{73}). \tag{7.42}$$

Since we must have $\xi = 0$ as $\rho \to \infty$, we must take the positive root $k \cong 3.77$. Hence

$$\chi(\rho) \xrightarrow[\rho \to \infty]{} \frac{144}{\rho^3} + \frac{B}{\rho^{3.77}}, \tag{7.43}$$

where B is an arbitrary constant.

To the same accuracy as (7.43) we can write alternatively

$$\chi(\rho) \xrightarrow[\rho \to \infty]{} \frac{144}{\rho^3 \left(1 + \dfrac{C}{\rho^{0.77}}\right)^n}, \tag{7.44}$$

where C and n are for the moment undetermined constants.

We now choose n so that $\chi(0)$ shall be finite; this requires $3 - 0.77n = 0$, or $n = 3.90$. Next we find C from the condition $\chi(0) = 1$; this gives $144/C^{3.90} = 1$, whence $C = (144)^{0.77/3} = (12^{2/3})^{0.77}$. Thus, finally,

$$\chi(\rho) \cong \left\{ 1 + \left(\frac{\rho}{12^{2/3}} \right)^{0.77} \right\}^{-3.90} . \qquad (7.45)$$

This approximate solution of A. Sommerfeld is in satisfactory agreement with the exact solution found by numerical integration; see Table 1.

Table 1.

ρ	χ_{approx}	χ_{exact}	ρ	χ_{approx}	χ_{exact}
0	1	1	2	0.22	0.24
0.3	0.67	0.72	3	0.14	0.16
0.5	0.56	0.61	5	0.072	0.079
1	0.38	0.42	∞	0	0

PROBLEMS

Problem 1. Obtain the expression (7.31) for the overlap integral by means of direct derivation.

Problem 2. Consider similarly the lowest single-particle excited state in the above example. Calculate the exact and self-consistent energies and the wave functions and the overlap integral.

Problem 3. Generalize the example considered in the text to the case of two fermions (the so-called *Hartree-Fock approximation*). Derive the approximate self-consistent wave function as the determinant composed of single-particle states.

Problem 4. Investigate the one-dimensional system of two coupled oscillators with the Hamiltonian

$$\hat{H} = -\frac{1}{2} \left(\frac{\partial^2}{\partial x_1^2} + \frac{\partial^2}{\partial x_2^2} \right) + \frac{1}{2}(x_1^2 + x_2^2) + \frac{1}{2} k(x_1 - x_2)^2 .$$

Compare the exact solution with the result of the Hartree approximation for the energy of the ground state.

Problem 5. Solve the previous problem in the case of three coupled oscillators, i.e. for the Hamiltonian

$$\hat{H} = -\frac{1}{2}\left(\frac{\partial^2}{\partial x_1^2} + \frac{\partial^2}{\partial x_2^2} + \frac{\partial^2}{\partial x_3^2}\right) + \frac{1}{2}(x_1^2 + x_2^2 + x_3^2)$$

$$+ \frac{1}{2} k[(x_1 - x_2)^2 + (x_1 - x_3)^2 + (x_2 - x_3)^2].$$

Compare the exact energy of the ground state with the result of the Hartree approximation as a function of the coupling parameter k.

Problem 6. Using the variational principle, derive the energy of the ground state of the helium atom. The single-particle wave functions of each of two electrons are chosen in hydrogen-like form

$$\psi(r) = \sqrt{\frac{Z^3}{\pi}} \exp(-Zr).$$

Here Z is the variational parameter (the effective charge of the atomic core). Answer: $Z = 27/16$; $E = -2.85$ a.u. (the exact experimental value is -2.90 a.u.!).

Problem 7. Obtain the self-consistent kinetic equations of A. Vlasov for the distribution functions in a collisionless plasma containing single-charged ions (i) and electrons (e):

$$\frac{\partial f_e}{\partial t} + v\frac{\partial f_e}{\partial r} - e\left\{E + \frac{1}{c}[v, H]\right\}\frac{\partial f_e}{\partial p} = 0;$$

$$\frac{\partial f_i}{\partial t} + v\frac{\partial f_i}{\partial r} + e\left(E + \frac{1}{c}[v, H]\right)\frac{\partial f_i}{\partial p} = 0.$$

Here **E**, **H** are the strengths of the electric and magnetic fields, respectively. They are determined from Maxwell's equations

$$\text{rot}\mathbf{E} = -\frac{1}{c}\frac{\partial \mathbf{H}}{\partial t}; \qquad\qquad \text{div}\mathbf{H} = 0;$$

$$\text{rot}\mathbf{H} = -\frac{1}{c}\frac{\partial \mathbf{E}}{\partial t} + \frac{4\pi}{c}\mathbf{j}; \qquad\qquad \text{div}\mathbf{E} = 4\pi\rho.$$

Here ρ, **j** are the averaged densities of charges and currents. They are derived via the distribution functions:

$$\rho = e\int (f_i - f_e)d^3\mathbf{p}; \qquad \mathbf{j} = e\int (f_i - f_e)\mathbf{v}d^3\mathbf{p}.$$

Chapter 8

Soliton Solutions

8.1 COUPLED PENDULA

This chapter is devoted to derivation of typical soliton solutions of nonlinear partial differential equations. The word 'soliton' means some solitary wave which is moving in space and changing its form with time. As in previous chapters, instead of the general theory we consider one simple example. Namely, we consider a system of identical mathematical pendula whose pivots are placed equidistant along the horizontal axis. All pivots of the pendula are connected to each other by elastic springs. The pendula can oscillate in the plane which is perpendicular to the connecting axis.

If we take φ_i to be the angle the ith pendulum makes with the vertical, we can write Newton's equation of motion for the system as

$$ml^2 \frac{d^2 \varphi_i}{dt^2} = -mgl \sin \varphi_i + k(\varphi_{i+1} - \varphi_i) + k(\varphi_{i-1} - \varphi_i). \quad (8.1)$$

Here m is the mass of each pendulum, l is its length, g is the free fall acceleration, and k is the torque constant of each spring.

If we take d to be the distance between the neighbouring pendula, and further $d \to 0$, then we obtain the continuous distribution of pendula along the horizontal axis:

$$(\varphi_{i+1} - \varphi_i) - (\varphi_{i-1} - \varphi_i) = d^2 \frac{\partial^2 \varphi}{\partial x^2}. \quad (8.2)$$

125

Here x is the coordinate along the horizontal axis where the pivots of the pendula are placed.

Substituting Eq.(8.2) into Eq.(8.1) we obtain a nonlinear partial differential equation for the deviation angle φ of the continuously distributed pendula:

$$\frac{\partial^2 \varphi}{\partial t^2} = -\omega^2 \sin \varphi + c^2 \frac{\partial^2 \varphi}{\partial x^2}. \tag{8.3}$$

Here the notation

$$\omega \equiv \sqrt{\frac{g}{l}}$$

is introduced: it is the frequency of small harmonic oscillations of the pendulum in the gravity field, and

$$c \equiv \frac{d}{l}\sqrt{\frac{k}{m}}$$

is the velocity of propagation of linear sound waves along the horizontal axis as $g \to 0$ i.e. as $\omega \to 0$. Replacing the discrete distribution of pendula by a continuous distribution is correct when the sound wavelength is large compared to the distance between the neighbouring pendula.

Equation (8.3) is called the *Sine-Gordon equation*. Our goal is to find its soliton solutions, i.e. solutions which describe solitary pulses. We are not interested in solutions in the form of simple sound waves or in solutions of intermediate type between sound and solitary waves (of course, they also exist).

It should be noted that besides the above mechanical example, the Sine-Gordon equation (8.3) describes various other physical models, for example, the motion of defects (in particular, of empty lattice sites) in crystal lattice of solids according to Frenkel theory.

8.2 THE SEARCH FOR SOLITON SOLUTIONS

In order to find soliton solutions of Eq.(8.3) it is convenient to introduce the new dependent variable u instead of the deviation angle φ :

$$\varphi = \arctan u. \tag{8.4}$$

Differentiating this relation with respect to x, we obtain

$$\frac{\partial \varphi}{\partial x} = \frac{4}{1 + u^2} \frac{\partial u}{\partial x}.$$

A second differentiation produces the result

$$\frac{\partial^2 \varphi}{\partial x^2} = \frac{4}{1 + u^2} \frac{\partial^2 u}{\partial x^2} - \frac{8u}{(1 + u^2)^2} \left(\frac{\partial u}{\partial x} \right)^2. \tag{8.5}$$

Analogously we obtain the second derivative over time:

$$\frac{\partial^2 \varphi}{\partial t^2} = \frac{4}{1 + u^2} \frac{\partial^2 u}{\partial t^2} - \frac{8u}{(1 + u^2)^2} \left(\frac{\partial u}{\partial t} \right)^2. \tag{8.6}$$

We also use the relation

$$\sin \varphi = \sin(4 \arctan u) = \frac{4u(1 - u^2)}{(1 + u^2)^2}. \tag{8.7}$$

Substituting Eqs.(8.5 – 8.7) into Eq.(8.3) we obtain the following partial differential equation for the function $u(x,t)$:

$$
\frac{4}{1+u^2}\left(c^2\frac{\partial^2 u}{\partial x^2}-\frac{\partial^2 u}{\partial t^2}\right)-\frac{8u}{(1+u^2)^2}\left\{c^2\left(\frac{\partial u}{\partial x}\right)^2-\left(\frac{\partial u}{\partial t}\right)^2\right\}
$$

$$
=\frac{4\omega^2 u(1-u^2)}{(1+u^2)^2}.
$$

After some cancellations this equation takes the final form

$$
(1+u^2)\left(c^2\frac{\partial^2 u}{\partial x^2}-\frac{\partial^2 u}{\partial t^2}\right)-2u\left\{c^2\left(\frac{\partial u}{\partial x}\right)^2-\left(\frac{\partial u}{\partial t}\right)^2\right\}
\tag{8.8}
$$

$$
=\omega^2 u(1-u^2).
$$

Let us seek a solution of this partial differential equation in the special form which is specific for soliton solutions:

$$
u(x,t)=\frac{f(x)}{g(t)}.
\tag{8.9}
$$

Differentiating with respect to the variable x we obtain

$$
\frac{\partial u}{\partial x}=\frac{f'}{g},\qquad\qquad\frac{\partial^2 u}{\partial x^2}=\frac{f''}{g};
\tag{8.10a}
$$

analogously, differentiating with respect to the variable t we obtain

$$
\frac{\partial u}{\partial t}=-\frac{f}{g^2}g',\qquad\qquad\frac{\partial^2 u}{\partial t^2}=-\frac{f}{g^2}g''+\frac{2f}{g^3}\left(g'\right)^2.
\tag{8.10b}
$$

Here primes mean derivatives with respect to the independent variable for the function (f or g).

Substituting Eq.(8.10) into Eq.(8.8) and multiplying by g^3 we obtain

$$(f^2 + g^2)\left[c^2 f'' - \frac{2f}{g^2} g'^2 + \frac{f}{g} g''\right] - 2f\left[c^2 f'^2 - \frac{f^2}{g^2} g'^2\right]$$

$$= \omega^2 f(g^2 - f^2).$$

After cancellation of two terms and rearrangement of the other terms we find

$$c^2 f^2 f'' - 2c^2 f \cdot f'^2 + \omega^2 f^3 + \frac{f^3}{g} g''$$

(8.11)

$$+ c^2 g^2 f'' - \omega^2 f g^2 + f g \cdot g'' - 2f \cdot g'^2 = 0.$$

We shall investigate various solutions of Eq.(8.11).

8.3 ONE SOLITON

It follows from the structure of Eq.(8.11) that exponential functions are the solutions of this equation for a definite choice of parameters of functions. Thus, let us seek functions $f(x)$ and $g(t)$ in the exponential form:

$$f(x) = \exp(x/a), \qquad\qquad g(t) = \exp(Vt/a), \qquad\qquad (8.12)$$

where a and V are some constants.

Substitution Eq.(8.12) into Eq.(8.11) produces two algebraic equations which make the two brackets equal to zero in Eq.(8.11):

$$(c^2 - \omega^2 a^2 - V^2)\exp(3x/a) = 0;$$

$$(c^2 - \omega^2 a^2 - V^2)\exp(x/a - 2Vt/a) = 0.$$

Obviously both equations are equivalent to each other. Hence, we obtain a connection between the parameters a and V:

$$a = \frac{1}{\omega}\sqrt{c^2 - V^2}. \qquad (8.13)$$

Substituting Eq.(8.12) into Eq.(8.9) and further Eq.(8.9) into Eq.(8.4), we obtain the solution in the form of a *solitary wave* (soliton):

$$\varphi = 4 \arctan\left[\exp\frac{x - Vt}{a}\right]. \qquad (8.14)$$

It is seen that this soliton is propagating with velocity V (the value of this velocity can be both positive and negative). The angular velocity of oscillations of a pendulum can be found from Eq.(8.14):

$$\frac{\partial\varphi}{\partial t} = -\frac{(2V/a)}{\cosh[(x - Vt)/a]}. \qquad (8.15)$$

The dependence of the angular velocity on the coordinate x for the fixed time instant t is shown in Fig. 12 ($t = 0$ is chosen). The parameter a determines the width of the soliton. According to Eq.(8.13) the width of the soliton decreases with increasing velocity V. According to Eq.(8.15) the amplitude of the angular velocity of the soliton (that is, its maximum value $2V/a$) increases with increasing velocity of propagation V. Thus, a narrow soliton has more power.

It should be noted that the velocity of propagation of a solitary wave V is small compared to the speed of sound wave c (the latter propagates along the strings in the absence of gravity ($g \rightarrow 0$, i.e. $\omega \rightarrow 0$).

8.4 TWO SOLITONS

In addition to the solution of Eq.(8.11) in exponential form described in the previous section, we can also try solutions in hyperbolic form:

$$f(x) = b\sinh(x/a); \qquad\qquad g(t) = \cosh(Vt/a). \qquad (8.16)$$

Here a, b and V are constants. Relations between these constants are obtained by substitution of Eq.(8.16) into Eq.(8.11).

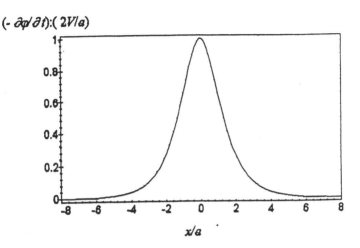

Figure 12. Single-soliton solution (8.15) of the Sine-Gordon equation.

In the process of this substitution let us collect the terms which are proportional to $\sinh^3(x/a)$:

$$c^2 b^3 - 2c^2 b^3 + \omega^2 a^2 b^3 + V^3 b^3 = 0.$$

Further we collect all terms which are proportional to $\sinh(x/a)\cdot\cosh(Vt/a)$:

$$c^2 b - \omega^2 ba^2 + bV^2 - 2bV^2 = 0.$$

It is seen that both equations are identical to each other. We again obtain Eq.(8.13) as a connection between the constants a and V.
    Finally let us write the terms which are proportional to $\sinh(x/a)$:

$$- 2c^2 b^3 + 2bV^2 = 0.$$

It follows from this relation that $b = V/c$. Hence, the second solution of the Sine-Gordon equation is of the form:

$$\varphi = 4\arctan\left\{\frac{V}{c}\frac{\sinh(x/a)}{\cosh(Vt/a)}\right\}. \tag{8.17}$$

The angular velocity of oscillation of the pendulum according to Eq.(8.17) is given by the expression

$$\frac{\partial\varphi}{\partial t} = -\frac{4V^2}{ac}\frac{\sinh(x/a)\cdot\sinh(Vt/a)}{\left[\cosh^2(Vt/a)+(V/c)^2\sinh^2(x/a)\right]}. \tag{8.18}$$

This function is shown in Fig. 13 for two times: the first time is large, while the second time is small compared to the typical time a/V.

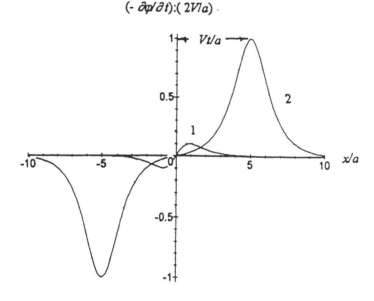

Figure 13. Two-soliton solution (8.18) of the Sine-Gordon equation; 1 – the case of small time; 2 – the case of large time (compared to a/V).

It is seen that the angular velocity is an odd function of the coordinate x for the two-soliton solution, while it is an even function for the single soliton (see the previous section). The two-soliton solution contains two solitary waves for all times: the first has positive amplitude, while the second has negative amplitude.

Minimal and maximal values of the angular velocity as a function of the coordinate x are determined from the condition that the derivative of Eq.(8.18) over x is equal to zero (it is more convenient to differentiate the inverse function):

$$\frac{V^2}{ac^2} \cosh(x/a) - \frac{\cos^2(Vt/a)}{a \sinh^2(x/a)} \cosh(x/a) = 0.$$

The positions of the maximal and minimal values are found from the equation:

$$\sinh(x/a) = \pm \frac{c}{V} \cosh(Vt/a). \tag{8.19}$$

The minimum distance between solitons occurs when $t=0$; the corresponding coordinate of each extremum is determined from the equation

$$\sinh(x_0/a) = \pm c/V.$$

Substituting Eq.(8.19) into Eq.(8.18) we obtain the soliton amplitude (i.e., the maximum value of the angular velocity of the pendulum):

$$\left. \frac{\partial \varphi}{\partial t} \right|_{max} = \frac{2V}{a} \tanh\left(\frac{Vt}{a} \right)$$

This amplitude has a maximum value of $2V/a$ for infinite times $t \to \pm\infty$.

Thus, the solitons are moving towards each other while their amplitudes diminish. Then they begin to repel each other and approach infinity. At $t=0$ the amplitude of each soliton is equal to zero, but they do not coincide. We can conclude that this solution describes the scattering of one soliton on the other soliton without their annihilation.

8.5 SOLITON + ANTISOLITON

The next solitary solution of Eq.(8.8) can be obtained from the solution found in the previous section, if we introduce new independent variables

$$x' \equiv Vt, \qquad\qquad t' \equiv x/V.$$

Let us also introduce new parameters

$$c' \equiv V^2/c, \qquad\qquad \omega' \equiv i\omega V/c.$$

Then we find that the parameter a does not change under these transformations since

$$a' = \frac{\sqrt{c'^2 - V^2}}{\omega'} = \frac{(iV/c)\sqrt{c^2 - V^2}}{(i\omega V/c)} = a.$$

Equation (8.8) takes the following form after these transformations:

$$(1 + u^2)\left\{ \frac{V^4}{c^2\,V^2} \frac{\partial^2 u}{\partial t'^2} - V^2 \frac{\partial^2 u}{\partial x'^2} \right\}$$

$$- 2u\left\{ \frac{V^4}{c^2}\left(\frac{\partial u}{V\partial t'} \right)^2 - V^2\left(\frac{\partial u}{\partial x'} \right)^2 \right\} = -(\omega V/c)^2 u(1 - u^2).$$

It is seen that after dividing by the factor $-(V/c)^2$ we again obtain Eq.(8.8). Hence, according to Eq.(8.17) the solitary solution of the Sine-Gordon equation is written in the form

$$\varphi = 4 \arctan u = 4 \arctan\left\{ \frac{V}{c'} \frac{\sinh(x'/a)}{\cosh(Vt'/a)} \right\}.$$

After returning to the old variables we obtain

$$\varphi = 4 \arctan\left\{-\frac{c}{V}\frac{\sinh(Vt\,/\,a)}{\cosh(x\,/\,a)}\right\}. \tag{8.20}$$

Differentiating this equation with respect to time we find the angular velocity of the oscillations of the pendulum:

$$\frac{\partial \varphi}{\partial t} = \frac{4c}{a}\frac{\cosh(x\,/\,a)\cdot\cosh(Vt\,/\,a)}{\{\cosh^2(x\,/\,a)+(c\,/\,V)^2\sinh^2(Vt\,/\,a)\}}. \tag{8.21}$$

This quantity is an even function of the coordinate x. The dependence of the angular velocity on x for the fixed time moment t is shown in Fig. 14.

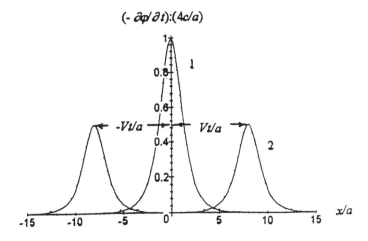

Figure 14. Soliton + antisoliton solution of the Sine-Gordon equation; 1 – the case of small time; 2 – the case of large time.

Let us determine the positions of the maxima in Fig.14. Differentiating the function which is inverse with respect to Eq.(8.21) we obtain the equation for the extremum coordinates:

$$\sinh(x / a) - \frac{c^2}{V^2} \sinh^2(Vt / a) \frac{\sinh(x / a)}{\cosh^2(x / a)} = 0.$$

Simplifying this equation we find three equations:

$$\cosh(x / a) = \pm(c / V) \sinh(Vt / a);$$

$$x = 0.$$

(8.22)

If the time t is small so that the condition

$$\left|\sinh(Vt / a)\right| < (V / c)$$

is fulfilled, then there is only one maximum (at $x = 0$, see Fig. 14). Under the opposite sign of the condition, i.e.

$$\left|\sinh(Vt / a)\right| > (V / c)$$

we have two solitary waves (see also Fig. 14). The amplitudes of the angular velocity can be obtained by substituting Eq.(8.22) into Eq.(8.21):

$$\left.\frac{\partial \varphi}{\partial t}\right|_{max} = \frac{2V}{a} \coth\left(\frac{Vt}{a}\right)$$

(8.23)

Thus, two solitons collide with each other, their amplitudes increasing according to Eq.(8.23). Then they annihilate each other into one soliton (therefore this solution is called a *soliton + antisoliton* solution).

After some time the new soliton decays again into two solitons which go to infinity, and their amplitudes decrease.

It should be noted that in the case of defects in a crystal lattice of solids such a soliton represents a site (no atom is in the lattice cell) while the antisoliton represents the additional atom in the lattice cell.

8.6 BREATHER SOLITON

Once more type of solution of the Sine-Gordon equation can be obtained
from Eq.(8.20) by means of the substitution $V \rightarrow iV$:

$$\varphi = 4 \arctan \left\{ -\frac{c \, \sin(Vt / a)}{V \, \cos(x / a)} \right\}. \tag{8.24}$$

According to Eq.(8.13) we now have the next expression for the parameter
a:

$$a = \frac{\sqrt{c^2 + V^2}}{\omega}. \tag{8.25}$$

In this case (unlike the previous cases) the velocity of propagation of the
soliton V can be both smaller than and larger than the sound speed c.
   The angular velocity of the pendulum can be obtained from Eq.(8.22)
by means of the same substitution $V \rightarrow iV$:

$$\frac{\partial \varphi}{\partial t} = \frac{4c}{a} \frac{\cosh(x / a) \cdot \cos(Vt / a)}{\left[\cosh^2 (x / a) + (c / V)^2 \sin^2 (Vt / a) \right]}. \tag{8.26}$$

Analogously to the previous case this angular velocity is an even function of
the coordinate x.
   The extremum coordinates can be found also from Eq.(8.22) by mean
of the same substitution $V \rightarrow iV$:

$$\cosh(x / a) = \pm(c / V) \sin(Vt / a);$$

$$x = 0. \tag{8.27}$$

Eq.(8.27) has no solutions under the condition $V > c$, and then the
dependence (8.26) describes one soliton with a maximum at $x = 0$, which is
of the form of a standing wave (Fig. 15), unlike the previous cases where all
solitons represent travelling waves.

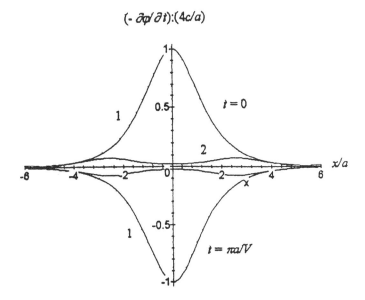

Figure 15. Breather soliton, Eq. (8.26), of the Sine-Gordon equation; 1 – the case $V > c$; 2 – the case $V < c$.

According to Eq.(8.26) the maximum is replaced periodically by the minimum (the latter corresponds to the negative values of the angular velocity of the pendulum). The temporal period is equal to

$$T = \frac{2\pi a}{V}.$$

The parameter a determines the width of the soliton analogously to the previous cases. Such a soliton is called a *bion* , or *breather* .

In the opposite case of $V < c$ two maxima appear periodically (see Fig. 15). They are described by Eq.(8.27). Then they become a single standing solitary wave which is analogous to that shown in Fig. 15 for the case $V > c$. In one half of the period the picture repeats, but below the abscissa axis in Fig. 15 since the angular velocity of the pendulum is negative.

PROBLEMS

Problem 1. Obtain the solitary solution of the equation which describes nonlinear diffusion:

$$\frac{\partial u}{\partial t} = \frac{\partial^2 u}{\partial x^2} + u(1 - u)(u - a).$$

Answer:

$$u = \frac{1}{1 + \exp\left(\dfrac{x - Vt}{\sqrt{2}}\right)},$$

where the velocity of the soliton is equal to

$$V = \frac{1}{\sqrt{2}}(1 - 2a),$$

and a is an arbitrary parameter.

Problem 2. Obtain a solitary solution of the equation for the so-called *Toda chain*:

$$\frac{\partial^2 u_n}{\partial t^2} = \exp(u_{n+1} - u_n) - \exp(u_n - u_{n-1}).$$

Here u_n is the coordinate of the nth particle in the chain.
Answer:

$$u_n(t) = \ln\left\{ \frac{1 + \exp[2(na \pm t \sinh a)]}{1 + \exp[2(na + a \pm t \sinh a)]} \right\}.$$

Here a is an arbitrary parameter.

Problem 3. Find the solitary solution for the *Korteweg – de Vries* equation

$$\frac{\partial v}{\partial t} + (u + v)\frac{\partial v}{\partial x} + \frac{uh^2}{6}\frac{\partial^3 v}{\partial x^3} = 0.$$

Obtain also the harmonic solution of this equation.
Answer:

$$v(x, t) = \frac{3V}{\cosh^2\left\{\sqrt{\frac{3V}{2u}}\ \frac{x - ut - Vt}{h}\right\}}.$$

This is a soliton with a velocity of propagation $u + V$. The velocity of the harmonic travelling wave is equal to u.

Problem 4. The one-dimensional wave motion of ions in a classical quasi-neutral plasma along the x-axis is approximated by the one-dimensional Euler equation (for single-charged ions, for the sake of simplicity)

$$\frac{\partial v}{\partial t} + v\frac{\partial v}{\partial x} = -\frac{e}{M}\frac{\partial \varphi}{\partial x},$$

where v is the velocity of an ion component, M, e are the mass and the charge of an ion, respectively, and φ is the potential of the electric field in the plasma. The concentration of ions N_i satisfies the one-dimensional equation of continuity

$$\frac{\partial N_i}{\partial t} + \frac{\partial}{\partial x}(N_i v) = 0.$$

The third equation is the one-dimensional Poisson equation for the electric potential

$$\frac{\partial^2 \varphi}{\partial x^2} = -4\pi e(N_i - N_e).$$

The concentration of electrons in this potential is determined by the known expression (for a weak electric field)

$$N_e = N_0\left\{1 + \frac{e\varphi}{T} - \frac{4}{3\sqrt{\pi}}\left(\frac{e\varphi}{T}\right)^{3/2}\right\}.$$

Here T is the electron temperature, and N_0 is the equilibrium concentration. The ion temperature is assumed to be small compared to the electron temperature. The last term in the right side of this equation produces a soliton wave.

Obtain this expression for the electron concentration using the Boltzmann distribution of electrons in the electric field and the adiabatic invariant (electrons follow adiabatically the variations of a weak electric field):

$$\int_a^b p_x dx = \text{const};$$

$$\varepsilon = \frac{p_x^2(x)}{2m} - e\varphi(x) = \text{const}.$$

Here ε is the total electron energy, and m is its mass.

Obtain the profile of a weak soliton wave for the weak electric potential in the form

$$\varphi = \varphi_m \cosh^{-4}\left[\frac{(x - ut)}{\sqrt{15a_e}}\left(\frac{e\varphi_m}{\pi T}\right)^{1/4}\right].$$

Here $\varphi_m \ll T$ is the small amplitude of the soliton,

$$a_e = \sqrt{\frac{T}{4\pi N_0 e^2}}$$

is the Debye radius for the electron component, and u is the velocity of the soliton.

Obtain also the expression for the velocity of the soliton

$$u = \sqrt{\frac{T}{M}}\left[1 + \frac{16}{15}\sqrt{\frac{e\varphi_m}{\pi T}}\right]^{1/2}.$$

Chapter 9

Dynamic Chaos

9.1 A PARTICLE IN A POTENTIAL BOX WITH VIBRATING WALL

If the perturbation of a classical system is random, then, of course, the motion of the system becomes stochastic (see Chapter 11). But it is not so obvious that the motion of a simple system with small degrees of freedom can be stochastic under the action of a regular perturbation, for example, of a sinusoidal force. We consider the mathematics of this problem using a simple mechanical example in order to avoid mathematical difficulties connected with the properties of the unperturbed system.

Consider the stochastic heating of a gas in a box with small dimensions. One of the walls of this box is supposed to be vibrating. We assume that the mean free path is large compared to the dimensions of the box (this is the so-called *Knudsen gas*). Let us simplify the problem by considering a separate particle of the gas in a one-dimensional infinite potential well. One of the walls (we suppose that it is the right wall) oscillates according to a harmonic law so that the velocity of this wall is

$$V(t) = V_0 \sin \omega t.$$

The amplitude of the velocity V_0 is supposed to be small compared to the velocity of the gas molecule. The latter will be noted by v. The width l of the well is supposed to be large compared to the amplitude of vibration of the right wall a, i.e.

$$l \gg a = V_0 / \omega.$$

This model is depicted in Fig. 16.

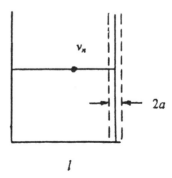

Figure 16. Stochastic heating of a one-dimensional gas in a box of the length l: the right wall of the box oscillates according to a sine-function law.

Let us write the energy conservation law at the elastic shock of a molecule upon the vibrating wall:

$$v_{n+1} = v_n + 2V_0 \sin(\omega t_n). \qquad (9.1)$$

Here v_n is the velocity of the molecule before the nth shock, and v_{n+1} is the velocity of this molecule before the next $(n + 1)$th shock, t_n is the time instant of the nth shock, and finally ω is the frequency of oscillation of the right wall. The left wall is supposed to be immovable, for the sake of simplicity.

Since the motion of the molecule between shocks is free, we can derive the time between the consequent shocks upon the right wall (see Fig. 16):

$$t_{n+1} - t_n = \frac{2l}{v_{n+1}}. \qquad (9.2)$$

Introducing the phase $\varphi \equiv \omega t$, we obtain from Eqs.(9.1) and (9.2) the so-called *Poincare transformation*:

$$v_{n+1} = v_n + 2V_0 \sin \varphi_n;$$

$$\varphi_{n+1} = \varphi_n + \frac{2l\omega}{v_{n+1}} \qquad \mathrm{mod}(2\pi). \qquad (9.3)$$

Here we take into account that the phase is determined in the interval $[0,2\pi]$. Our mathematical goal is to investigate the solutions of the system (9.3).

9.2 APPEARANCE OF DYNAMICAL CHAOS

We are interested first of all in how the phase φ_n depends on the number n. If the variation of this phase after each step is small compared to 2π, then it is clear that the dependence of the phase on time is regular. However, if the variation of the phase is of the order of 2π (this variation cannot be larger than 2π according to the definition of the phase) then this does not mean that the motion of the system will be stochastic. Indeed, the variation of the phase can be large, but if it is almost the same on the next step, then the variation of the molecular velocity determined from the first of Eqs.(9.3) will obviously be regular.

Chaotic variation of the velocity appears only when the variation of the phase is not only large, but also *irregular* on each step. From the mathematical point of view we can express this statement by introducing the so-called *coefficient of phase stretching* :

$$K \equiv \frac{(\varphi_{n+1} - \varphi_n) - (\varphi_n - \varphi_{n-1})}{(\varphi_n - \varphi_{n-1})}. \qquad (9.4)$$

The motion of the system is regular if the value of this coefficient is small compared to 1. However, if the value of K exceeds 1, then the motion will be stochastic, because the molecular velocity on the next step strongly differs from the molecular velocity on the previous step, and this difference is of an irregular character.

In order to apply the condition (9.4) to our example we rewrite the condition for stochastic motion $K > 1$ in the following manner: the denominator of Eq.(9.4) is replaced by 1, because in the case when it is small compared to 1 the motion will be regular (see the above discussion). This denominator cannot be larger than 1 according to the above definitions. We use the second of Eqs.(9.3) for the phase differences which are contained in the numerator of Eq.(9.4). Thus, we obtain

$$K = 2l\omega\left(\frac{1}{v_{n+1}} - \frac{1}{v_n}\right) \approx \frac{2l\omega V_0}{v^2}. \qquad (9.5)$$

We have also used here the first of Eqs.(9.3): the molecular velocity varies irregularly, but its variation is small during each step compared to the velocity itself. This can be explained by the fact that the velocity of the vibrating wall is small compared to the molecular velocity, and therefore the molecular velocity varies by a small quantity during each collision with the wall.

9.3 DIFFUSION PROCESSES

If the conditions for stochastic motion are fulfilled, then it is interesting to understand how the molecular velocity changes during a large number of collisions with the wall: is it increasing or decreasing on the average? We show that an *increase* of the velocity takes place; we find also the law of this averaged increase with time. Firstly we introduce the *mean molecular velocity* $<v>$ by the relation

$$v - <v> \equiv v_{n+1} - v_n.$$

Further we take the square of the first of Eqs.(9.3) and average it over a large number of collisions with the wall:

$$\left\langle (v - <v>)(v - <v>) \right\rangle = 2V_0^2. \qquad (9.6)$$

Here we replace the square of the sine function of the phase by the factor 1/2 due to the random value of this phase in the right side of the first of Eqs.(9.3).

According to Eq.(9.2) we write the right side of Eq.(9.6) in the form

$$2V_0^2 \, \frac{dt}{(2l/v)}$$

where we have written the small interval between two consecutive collisions as dt. The left side of Eq.(9.6) represents the so-called *dispersion* of the velocity.

Omitting the averaging signs, we rewrite the dispersion in the form

$$<\Delta v^2> = d(v^2) = 2v dv.$$

Now the velocity v is a continuous function of time. Substituting the last two expressions into Eq.(9.6) we obtain

$$2dv = \frac{v_0^2}{l} dt .$$

The elementary solution of this differential equation is

$$v = v_0 + \frac{v_0^2}{2l} t . \qquad (9.7)$$

Thus, we find that the diffusion increases the molecular velocity with time according to a linear law of uniform acceleration.

We have seen that the molecular velocity varies chaotically at each collision, but we obtain an increase of this velocity on the average. When does this increase stop? The answer is: when the coefficient of phase stretching K becomes smaller than unity. Using Eq.(8.6) we find that the increase of velocity stops when this velocity achieves its maximum value

$$v = v_{max} \approx \sqrt{\frac{2l\omega}{v_0}} \gg v_0 . \qquad (9.8)$$

Equation (9.7) also determines the diffusion time for an increase of the velocity up to its maximum value, Eq.(9.8):

$$t_D = v_{max} \frac{2l}{v_0^2} = \frac{l}{\omega}\left(\frac{2l\omega}{v_0}\right)^{3/2} \gg \frac{1}{\omega} . \qquad (9.9)$$

From the physical point of view the reason for the acceleration of the molecule consists in the fact that the molecule which is moving towards the vibrational wall collides more often with this wall than the molecule which is moving away from the vibrational wall. Hence, an increase of the molecular velocity takes place more often than a decrease. At each collision with the wall the increase or decrease of the molecular velocity is small compared to its velocity. Therefore the average increase of the molecular velocity is a diffusion process.

9.4 THE FOKKER - PLANCK EQUATION

The diffusion approximation is valid if the investigated physical quantity changes a few times during small time intervals, but chaotically. According to the well known law of the one-dimensional diffusion process we have

$$< \Delta v^2 > = D \cdot \Delta t \qquad (9.10)$$

for the dispersion of the velocity during the time interval Δt. Here D is the diffusion coefficient. Then it follows from Eq.(9.6) that the expression for the diffusion coefficient is

$$D = \frac{V_0^2}{l} v. \qquad (9.11)$$

Here we have used the expression $\Delta t = 2l/v$ for the time interval between two consecutive collisions of the molecule upon the vibrating wall.

It is seen that the diffusion coefficient depends on the variable velocity v so that the diffusion equation (more exactly, the *Fokker-Planck equation*) is of the form

$$\frac{\partial F}{\partial t} = \frac{1}{2} \frac{\partial}{\partial v} \left(D(v) \frac{\partial F}{\partial v} \right) \qquad (9.12)$$

Here $F(v,t)$ is the distribution function of molecules depending upon their velocities; it depends also on time. This function is supposed to be normalized by 1 for each time instant.

Substituting Eq.(9.11) into Eq.(9.12), we derive the average molecular velocity again and we must obtain, of course, the above result, Eq.(9.7). Multiplying Eq.(9.12) by the molecular velocity v and integrating over this velocity, we find after double integration by parts:

$$\frac{\partial < v >}{\partial t} = \frac{V_0^2}{2l} \left\{ \int v \frac{\partial}{\partial v} \left(v \frac{\partial F}{\partial v} \right) dv \right\} = -\frac{V_0^2}{2l} \int v \frac{\partial F}{\partial v} dv = \frac{V_0^2}{2l}.$$

We again obtain the accelerated motion of the molecule on the average, and the uniform acceleration coincides with Eq.(9.7) as it should do.

9.5 REGULAR MOTION

If the time exceeds the diffusion time, determined by Eq.(9.9), then the distribution function of molecules upon their velocities $F(x,t)$ takes a stationary value. Molecules with velocities exceeding the estimate (9.8) are moving along regular trajectories. Let is find the characteristics of this motion in our example (see above).

If the molecular velocity satisfies the inequality $v_{max} < v$ then the right side of the second of Eqs.(9.3) becomes small compared to 1,   i.e. the variation of the phase during one collision becomes small. In this case the system of equations (9.3) can be replaced by differential equations. Dividing the first of Eqs.(9.3) by the second one, we find

$$\frac{dv}{dt} = \frac{V_0 \, v}{l} \sin \omega t . \tag{9.13}$$

Separating variables in this equation and integrating, we obtain

$$\ln v + \text{const} = - \frac{V_0}{l\omega} \cos \omega t,$$

and the regular solution takes the form

$$v(t) = u \cdot \exp\left[- \frac{V_0}{l\omega} \cos \omega t \right].$$

Here the integration constant $u \sim \omega \, l$. Due to inequality $V_0 << \omega \, l$ we can expand the exponential expression in a Taylor series and take into account the first and second terms only:

$$v(t) = u\left[1 - \frac{V_0}{l\omega} \cos \omega t \right].$$

Thus, the molecular velocity is practically constant, because the second term in the right side of the last expression is small compared to the first one. The small variation of the velocity is regular and it does not produce any diffusion increase of the velocity during a large number of periods of vibrations of the wall. This statement is valid also in the exact solution, when the Taylor expansion is incorrect.

In conclusion we can say that the molecule moves stochastically, and its average velocity increases according to a diffusion law. Then the diffusion stops, and only small regular variations of the molecular velocity produced by the vibrating wall take place.

It should be noted that it follows from numerical derivations that small regular variations of the molecular velocity stop after millions of shocks. Then *the velocity suddenly begins chaotically to decrease to zero*. This is a typical property of chaotic motion, not only for this problem, but also for all problems with dynamic chaos.

9.6 THE STANDARD MAP

It should be noted that the example considered above of a particle in a box with a vibrating wall is a quite general situation. The goal of this section is to show that the Poincare map of the type (9.3) as a rule produces dynamic chaos under the conditions when the coefficient of phase stretching is $K > 1$.

We again consider the case of coupled pendula which was investigated in the previous chapter. However, now we assume that the sound wavelength (the sound wave is propagating along the horizontal axis where the pivots are placed) is of the order of the distance between neighbouring pendula. Under these conditions we cannot consider the distribution of pendula as a continuous distribution. Thus, we consider the system of difference equations (8.1). Because of the difficulty of the problem we will restrict ourselves to the stationary case so that the left side of Eqs. (8.1) is equal to zero.

Thus, let us investigate the system of difference equations

$$\omega^2 \sin \varphi_n = \frac{k}{ml^2} [(\varphi_{n+1} - \varphi_n) - (\varphi_n - \varphi_{n-1})].$$

The following notation is introduced for the difference of phases:

$$I_n \equiv \varphi_n - \varphi_{n-1}.$$

Then the system of difference equations can be rewritten in the form of the so-called *standard* (or *universal*) *map*:

$$I_{n+1} = I_n + K \sin \varphi_n;$$

$$\varphi_{n+1} = \varphi_n + I_{n+1}, \qquad \mathrm{mod}(2\pi). \tag{9.14}$$

Here the notation

$$K = \frac{mgl}{k} \tag{9.15}$$

is introduced. This quantity represents the square of the ratio of the frequency of small oscillations of a pendulum to the frequency of eigen-vibrations of a spring which connects neighbouring pendula. Analogously to Eq.(9.3) the phase is placed in the interval $[0,2\pi]$ after each step.

The case of K is uninteresting, since a small deviation of one pendulum produces small deviations of other pendula (this case is described by perturbation theory). If the deviation of a pendulum is large then due to the smallness of values of I_n deviations of neighbouring pendula will be nearly equal to the deviation of the initial pendulum. This is also an uninteresting case.

More important is the case of $K > 1$, since then the phase stretching (see Eq.(9.4)) is of the order of the value

$$\frac{(\varphi_{n+1} - \varphi_n) - (\varphi_n - \varphi_{n-1})}{(\varphi_n - \varphi_{n-1})} \propto (I_{n+1} - I_n) \propto K. \tag{9.16}$$

Here we replace the denominator of the ratio by unity. Analogously to the above example of a molecule in a box with a vibrating wall we observe that the values of the phase φ vary not only strongly after each step, but also irregularly.

Due to the chaotic variation of the deviation angle $\varphi_n \sin\varphi_n$ takes all values in the interval $[-1,+1]$ quite uniformly. It should be noted that from the physical point of view the quantity I_n determines the degree of twisting of the nth string. Hence, it follows from the first of Eqs.(9.14) that

$$< \Delta I^2 >= \frac{1}{2} K^2. \tag{9.17}$$

Here the averaging is made over all pendula. The left side of Eq.(9.17) describes the dispersion of the quantity I on the one step, i.e. at the transition from one pendulum to the next pendulum. According to the second of Eqs.(9.14) the variation of the deviation angle on one step is equal to I. Hence, Eq.(9.17) can be rewritten in the form

$$< \Delta I^2 > = K^2 \frac{\Delta \varphi}{2I}. \tag{9.18}$$

Equation (9.18) describes the diffusion process for the variation of the quantity I with the variation of the phase φ. The diffusion coefficient is

$$D = \frac{K^2}{2I} \tag{9.19}$$

and it should be noted that it depends on the variable I.

Let us rewrite Eq.(9.18) in the differential form (analogously to such procedure in the analysis of a particle in a box with a vibrating wall, see above):

$$2I \mathrm{d}I = \frac{K^2}{2I} \mathrm{d}\varphi. \tag{9.20}$$

We can integrate this equation and find the dependence of the quantity I on the angle φ:

$$< I^3 (\varphi) > = < I^3 (\varphi_0) > + \frac{3K^2}{4} (\varphi - \varphi_0). \tag{9.21}$$

Thus, it is seen that the difference of the deviation angles of neighbouring pendula increases along the horizontal axis where the pivots are placed. Hence, neighbouring pendula oscillate irregularly with respect to each other at large deviation angles.

We conclude that if the frequency of small oscillations of the pendulum ω and the frequency of eigen-vibrations of the spring are of the same order of value, then by deviating one pendulum we obtain stochastic picture of deviations of pendula which is stationary in time.

It is necessary for the applicability of the diffusion approximation that the investigated quantity (in our case this is the number of revolutions of the spring $I/2\pi$) varies slightly compared to the value of the considered quantity, when we are going from one pendulum to the next pendulum.

Thus, this picture of diffusion is correct when the deviations of the pendula are large, and the springs connecting the pendula are twisting over a large number of revolutions, i.e. when $I \gg 1$. The number of revolutions differs by one, or two revolutions for the neighbouring pendula, i.e. relatively small compared to the value of I.

In the next section we discuss the appearance of dynamic chaos in more detail for the same example of stationary states of coupled pendula, but starting with small values of the coefficient of phase stretching, since in this case we can simply analyse the standard map.

9.7 PHASE PICTURE OF THE STANDARD MAP

We have said above that the dynamics of a classical system is regular at small values of the coefficient of phase stretching $K \ll 1$. In this case the system of equations (9.14) reduces from a difference to a differential system. Dividing the first of these equations by the second we obtain

$$\frac{I_{n+1} - I_n}{\varphi_{n+1} - \varphi_n} = \frac{dI}{d\varphi} = \frac{K \cdot \sin \varphi}{I}. \tag{9.22}$$

Its solution is of the simple form

$$I^2 = 2K(1 - \cos \varphi) + \text{const}. \tag{9.23}$$

Let us consider first the vicinity of immovable points of the map (9.14). According to Eq.(9.23) we have in the vicinity of the immovable point $I = 0$, $\varphi = \pi$:

$$I^2 + K^2(\varphi - \pi)^2 = \text{const}. \tag{9.24}$$

Thus, this immovable point is the elliptic point, and the trajectories around this point are stable (Fig. 17).

In the vicinity of the second immovable point $I = \varphi = 0$ we have, from Eq.(9.23):

V.P. KRAINOV

$$I^2 - K^2 \varphi^2 = \text{const.} \qquad (9.25)$$

This point is hyperbolic, and the trajectories around it are unstable (Fig. 17).

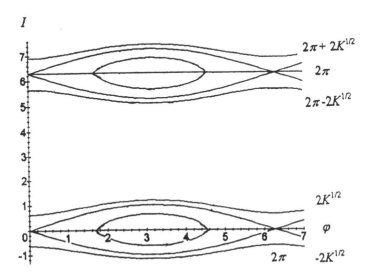

Figure 17. Phase picture of the standard map for small values of the coefficient of phase stretching K.

The phase picture of motion for this system is shown in Fig. 17. In the same figure we also show the separatrice which separates the finite trajectories from the infinite . Its equation can be obtained from Eq.(9.23) if we put const = 0:

$$I = \pm 2\sqrt{K} \sin(\varphi / 2). \qquad (9.26)$$

The separatrice goes through the unstable points $I = 0$, $\varphi = 0$, 2π, 4π,... of the hyperbolic type. Phase picture for the trajectories with immovable points $I = 2\pi$ is also shown in Fig. 17.

From the qualitative point of view, the destruction of regular motion along the considered trajectories appears when neighbouring separatrices overlap each other, i.e. under the condition

$$K > (\pi / 2)^2. \qquad (9.27)$$

The stochastic layer is formed in the vicinity of the separatrice. Then the particle moves according to the standard transformation (9.14) from the region inside one separatrice to the region inside the neighbouring separatrice, i.e. this particle moves along the vertical axes of I in Fig. 17. This corresponds to a diffusion process (see Sect.9.3) under the condition $I \gg 1$. The value of I increases on average when we are going from one pendulum to the next, i.e. the twisting of springs coupling pendula pivots increases along the horizontal axis where pivots are placed.

In this problem the value of the coefficient of phase stretching   (see Eq.(9.16)) does not depend on I . Hence, any limitations for diffusion increasing of the quantity I are absent. However, in the previous problem (a molecule in a box with a vibrating wall) such a dependence is present, and increasing of the molecular velocity was restricted.

Computer derivation of the standard map (9.14) allows us to reach the conclusion that the stochastic layers of the neighbouring separatrices overlap at $K > 1$, i.e. more quickly than according to Eq.(9.27).

Finally, if the coefficient of phase stretching is $K \gg 1$, then the region of stochastic motion fills in practically all of the phase picture on Fig. 17 except for small regions near the stable elliptic points $\varphi = \pi$, $I = 0$, 2π, 4π, ... (in this case the region of stochastic motion is called the *stochastic sea*).

It should be noted that we can restrict to the interval $[0, 2\pi]$ for the variable φ in the phase picture in all cases.

9.8 CHAOS IN TURBULENCE

In this section we consider a mechanism which qualitatively explains the onset of turbulent chaotic motion of a fluid with increasing Reynolds number. This mechanism was discussed by Feigenbaum.

Qualitatively, we can write the Navier – Stokes equation for a viscous fluid

$$\frac{\partial \mathbf{v}}{\partial t} + (\mathbf{v}\nabla)\mathbf{v} = -\frac{1}{\rho}\nabla p + v\Delta\mathbf{v}$$

($\mathbf{v}$, p, ρ are the velocity, pressure and density of the fluid, respectively; v is the kinematic viscosity) in the form

$$\frac{\partial v}{\partial t} = av^2 + bv + c. \tag{9.28}$$

Here we have assumed that the perturbation of the pressure is proportional to the velocity, which is correct for not too high pressures.

The differential equation (9.28) can be transformed to a difference equation. The left-hand side of (9.28) can be rewritten in the form

$$\frac{\partial v}{\partial t} = \frac{v(t+T) - v(t)}{T}.$$

Here T is the period of the turbulence vortex. Then we can rewrite (9.28) in the form of a difference equation

$$v(t+T) = \alpha v^2(t) + \beta v(t) + \gamma. \tag{9.29}$$

The linear term $\beta v(t)$ in the right-hand side of (9.29) can be omitted with no loss of generality by transforming to a reference frame moving with respect to the original frame with a certain constant velocity. Then (9.29) takes the form which qualitatively retains all of the main features of the original Navier – Stokes equation

$$v(t+T) = u - \frac{1}{u} \, \mathrm{Re} \cdot v^2(t). \tag{9.30}$$

Here $\mathrm{Re} = uR/v$ is the Reynolds number. Indeed, the ratio of the nonlinear term in the Navier – Stokes equation to the viscous term is of the order of the Reynolds number. Here, u and R are the characteristic velocity of the fluid, and the characteristic dimension of a body placed in the moving fluid, respectively. The minus sign in front of the quadratic term in this equation is chosen because otherwise the velocity would increase without bound as the motion repeated in time with period T ($u > 0$). Finally we can choose the system of units for the velocity of the fluid in which $u = 1$, and also we will measure the time in units of the period T, i.e. $t/T = n = 0, 1, 2,...$. Then (9.30) can be rewritten in the form of the so-called *logistic map*

$$v(n+1) = 1 - \mathrm{Re} \cdot v^2(n). \tag{9.31}$$

A strictly periodic solution of (9.31) implies that $v_1(n+1) = v_1(n)$. Therefore we obtain from (9.31) the quadratic equation

$$v_1 = 1 - \text{Re} \cdot v_1^2,$$

whose solution is

$$v_1 = \frac{\sqrt{1 + 4\,\text{Re}} - 1}{2\,\text{Re}}. \tag{9.32}$$

Here, we have chosen the positive root of the equation.

We now consider a small perturbation on this periodic solution: $v = v_1 + \Delta v$. Substituting this solution into (9.31) and keeping terms up to second order in the perturbation Δv, we find

$$\Delta v(n + 1) = -2\,\text{Re} \cdot v_1 \Delta v(n). \tag{9.33}$$

If $2\text{Re} \cdot v_1 < 1$, then according to (9.33) $|\Delta v(n+1)| < |\Delta v(n)|$ and the perturbation decreases in time.

Therefore periodic motion with the velocity (9.32) is stable. In the opposite case $2\text{Re} \cdot v_1 > 1$ the motion is unstable because the perturbation grows rapidly in time. The loss of stability occurs for values $\text{Re} = \text{Re}_1$, $v_1 = \langle v_1 \rangle$, where

$$2\,\text{Re}_1 \cdot \langle v_1 \rangle = 1. \tag{9.34}$$

Eliminating $\langle v_1 \rangle$ between the two equations (9.32) and (9.34), we find

$$\text{Re}_1 = 3/4. \tag{9.35}$$

Hence periodic motion with the velocity v_1 is stable for $\text{Re} < \text{Re}_1$ and is unstable for $\text{Re} > \text{Re}_1$.

According to (9.33) we have $\Delta v(n+1) = -\Delta v(n)$ on the stability boundary and therefore $\Delta v(n+2) = \Delta v(n)$. We see that motion with the velocity $\langle v_1 \rangle + \Delta v$ is periodic with period $2T$, i.e., with double the period of the original periodic motion. This assertion is correct not only when $\text{Re} = \text{Re}_1$, but for any Re. However, in the general case Δv is not an arbitrarily small perturbation, as above, but has a certain finite value corresponding to the velocity $v = \langle v_1 \rangle + \Delta v$ of the flow with period $2T$.

We find the condition for loss of stability of the periodic flow with period $2T$. We iterate (9.31) through another period:

$$v(n + 2) = 1 - \mathrm{Re} + 2\, \mathrm{Re}^2 v^2(n) - \mathrm{Re}^3 v^4(n). \qquad (9.36)$$

The last term on the right-hand side of (9.36) can be approximately neglected (it can be shown that $v < 1$ and therefore v^4 is small). We introduce the notation

$$v(n) = (1 - \mathrm{Re})\tilde{v}(n); \qquad \lambda = 2\,\mathrm{Re}^2\,(\mathrm{Re} - 1). \qquad (9.37)$$

Then (9.36) takes the form

$$\tilde{v}(n + 2) = 1 - \lambda \tilde{v}^2(n).$$

We see that this equation has the same form as (9.31) if we replace v:

$$v \rightarrow \tilde{v}$$

and Re by λ. Repeating the stability analysis given above, we obtain from (9.35) that the loss of stability of the periodic motion with period $2T$ occurs when $\lambda = \lambda_2 = \frac{3}{4}$. From (9.37) the corresponding Reynolds number is $\mathrm{Re}_2 = 1.23$.

Therefore periodic motion with the doubled period $2T$ is stable when $\mathrm{Re} < \mathrm{Re}_2 = 1.23$.

Performing another iteration of (9.31), it can be shown that the loss of stability of periodic motion with period $4T$ occurs when $\mu = \frac{3}{4}$, which is the analog of the condition (9.35), where

$$\mu = 2\lambda^2 (\lambda - 1)$$

[from (9.37)]. Therefore $\lambda = \lambda_3 = 1.23$ and we find from (9.37) that $\mathrm{Re}_3 = 1.34$.

Repeating this procedure an infinite number of times, we obtain that the total loss of stability of flow with all periods takes place for the critical Reynolds number $\mathrm{Re}_{cr} = \mathrm{Re}_\infty$. According to (9.37) this number is found from the equation

$$\mathrm{Re}_{cr} = 2\,\mathrm{Re}_{cr}^2\,(\mathrm{Re}_{cr} - 1). \qquad (9.38)$$

Hence we easily obtain

$$\mathrm{Re}_{cr} = \frac{1}{2}(1 + \sqrt{3}) \approx 1.37. \qquad (9.39)$$

The region $\mathrm{Re} > \mathrm{Re}_{cr}$ is the region we call turbulent flow. In this region all vortex motion is unstable and is rapidly destroyed a short time after its inception.

Note the rapid convergence of the sequence of Reynolds number Re_n with increasing n corresponding to a loss of stability of periodic motion with period $2^{n-1}T$.

Period doubling of the motion is one of the ways that turbulence originates.

9.9 THE METHOD OF RENORMALIZATION GROUPS

The above example of the transition to chaotic turbulent motion in dynamical systems can be generalized. Let us consider, instead of the concrete map (9.31), a sufficiently arbitrary Poincare map

$$x_{n+1} = f_0(x_n). \tag{9.40}$$

We assume without loss of generality only that the function f_0 is an even function of its argument (see the text between (9.29) and (9.30)) and this function has one maximum in the considered interval of the independent variable (if the function has a minimum then iterations of (9.40) do not converge: the quantity x increases up to infinity).

It follows from (9.40) that

$$x_{n+2} = f_0(f_0(x_n)).$$

Now we change the independent discrete variable $x_n \equiv y_n/a_0$. Then we can rewrite the previous equations in the form

$$y_{n+1} = a_0 f_0(y_n / a_0);$$
$$y_{n+2} = a_0 f_0(f_0(y_n / a_0) \equiv f_1(y_n). \tag{9.41}$$

Thus, the form of the second of Eqs.(9.41) is analogous to (9.40), but after two steps.

The still arbitrary parameter a_0 is chosen in this method to satisfy the equation

$$a_0 f_0(f_0(0)) = 1. \tag{9.42}$$

Next we repeat the previous operations once more. We introduce a new independent variable using the relation $y_n \equiv z_n/a_1$. The quantity a_1 is determined from the equation which is analogous to (9.42):

$$a_1 f_1 (f_1 (0)) = 1. \tag{9.43}$$

It follows from the second equation in (9.41) that

$$z_{n+2} = a_1 f_1 (z_n / a_1)$$

and

$$z_{n+4} = a_1 f_1 (f_1 (z_n / a_1)) \equiv f_2 (z_n). \tag{9.44}$$

The generalization of this procedure gives

$$f_{k+1} (u_n) = a_k f_k (f_k (u_n / a_k)). \tag{9.45}$$

Here the parameter a_k is determined from the generalization of (9.43):

$$a_k f_k (f_k (0)) = 1. \tag{9.46}$$

Now we can derive limit of these functions as $k \to \infty$. Thus, we find

$$f(u) = af(f(u / a)); \qquad af(f(0)) = 1. \tag{9.47}$$

Without loss of generality we can choose $f(0) = 1$ (see the transformation from (9.30) to (9.31)). Hence it follows from (9.47) that $a f(1) = 1$.

Let us seek the solution of the functional equation (9.47) in the form of a Taylor series in even powers (see above) of the independent variable:

$$f(u) = 1 + \sum_{m=1}^{\infty} b_m u^{2m}.$$

For example, let us restrict ourselves to only the first two terms of this series:

$$f(u) \approx 1 + b_1 u^2.$$

This is just the Poincare map (9.31). Then according to (9.47), $a = (1+b_1)^{-1}$. The first of Eqs. (9.47) takes the form

$$1 + b_1 u^2 \approx 1 + 2b_1^2 (1 + b_1) u^2.$$

Here we have neglected terms of the order of u^4. We find the quadratic equation $2b_1^2 + 2b_1 - 1 = 0$ whose solution is

$$b_1 = -(1 + \sqrt{3})/2 \approx -1.37.$$

Thus, we obtain the approximate solution of (9.47) in the form

$$f(u) \approx 1 - 1.37u^2.$$

This coincides with (9.31) for the critical values of the Reynolds number (9.39) as it should do.

If we take into account more terms in the Taylor expansion of $f(u)$, then numerical derivations of algebraic equations result in the universal function

$$f(u) = 1 - 1.528u^2 + 0.105u^4 + 0.027u^6 - .... \tag{9.48}$$

Then a more exact value for the parameter $a = 1/f(1)$ is $a \cong -2.50$. Hence, a more exact critical value of the Reynolds number in the previous section is according to (9.31)

$$\mathrm{Re}_{cr} = 1 - \frac{1}{a} \approx 1.40. \tag{9.49}$$

Rapid convergence of the Reynolds numbers to the critical value (9.49) can be investigated also in general form using the method of renormalization groups. Let us introduce the new parameter δ by the relation:

$$\mathrm{Re}_{cr} - \mathrm{Re}_k = \frac{1}{\delta^k}. \tag{9.50}$$

Our goal is to derive δ. According to (9.49) we can also write for the renormalization parameter the relation

$$\frac{1}{a_k} - \frac{1}{a} = \frac{1}{\delta^k},$$

so that δ is a *universal number* (the *Feigenbaum number*) independent of the form of the Poincare map.

It follows from (9.50) that

$$\text{Re}_{k+1} - \text{Re}_k = \frac{1}{\delta^k} - \frac{1}{\delta^{k+1}};$$

$$\text{Re}_{k+2} - \text{Re}_{k+1} = \frac{1}{\delta^{k+1}} - \frac{1}{\delta^{k+2}}.$$

Dividing these expressions by one another we obtain

$$\delta = \lim \frac{\text{Re}_{k+1} - \text{Re}_k}{\text{Re}_{k+2} - \text{Re}_{k+1}}, \qquad k \to \infty. \qquad (9.51)$$

Using $k = 1$, and the values of $\text{Re}_1 = 0.75$, $\text{Re}_2 = 1.23$ and $\text{Re}_3 = 1.34$ from the previous section, we find from (9.51) that $\delta = 4.36$. A more exact value (the Feigenbaum number) using $k \gg 1$ is

$$\delta \approx 4.67 \qquad (9.52)$$

Since this value is large compared to unity, convergence of the series is very rapid.

PROBLEMS

Problem 1. Using the computer, investigate the equation describing dynamics of increasing of population:

$$x_{n+1} = x_n[1 + r(1 - x_n)].$$

Here x_n is the population after n years, r is the growth parameter, and $x_n = 1$ is the restriction of this growth.

Show that for $r < 2$ the equilibrium value of $x_{n \to \infty} = 1$ is stable, but at $2 < r < 6^{1/2} \cong 2.449$ regular oscillations take place between two limiting values; one of these values is less than unity while other is larger than unity.

Problem 2. Under the conditions of the previous problem, show that for $r = 2.5$ the period of oscillations is doubled compared to the case of smaller values of the parameter r. This phenomenon is called *bifurcation* of the period.

Problem 3. Under the conditions of Problem 1 show that for $r > 2.570$ periodic processes disappear, and dynamical chaos appears in the values of the population x_n.

Problem 4. Consider the map

$$x_{n+1} = \{Kx_n\}, \qquad\qquad 0 < x < 1.$$

Here $\{...\}$ is the fractional part of the number, and $K > 1$ is a parameter. Show that this map produces a chain of chaotic numbers, and that the memory of the initial number x_0 is forgotten sufficiently quickly at large noninteger values of the parameter $K \gg 1$.

Problem 5. Consider the previous problem for the map

$$x_{n+1} = \{K \sin(\pi x_n)\}, \qquad\qquad 0 < x < 1.$$

Problem 6. Investigate the chaotic one-dimensional motion of a charged classical particle in the field of an electromagnetic wave packet :

$$\frac{d^2 x}{dt^2} = \frac{e}{m} F \cos(kx - \omega t) \sum_{n=-\infty}^{\infty} \cos[n(\Delta kx - \Delta \omega t)].$$

Consider the partial cases: (1) $\Delta k = 0$ (temporal wave packet), and (2) $\Delta \omega = 0$ (space wave packet).

Chapter 10

Prey – Predator Population Models

10.1 THE VOLTERRA – LOTKA MODEL OF INTERACTING POPULATIONS

This chapter is devoted to the mathematical model of interacting populations suggested by Volterra and Lotka. We begin with the simplest situation where there exist prey which can reproduce according to the well know *Malthus law*. Let $N_1(t)$ be their population as a function of time t. This population is restricted by predators which survive only by eating prey. We denote the population of predators as $N_2(t)$.

The system of two nonlinear differential equations describing these populations is of the obvious form

$$\frac{dN_1}{dt} = (k_1 - k_2 N_2)N_1;$$

$$\frac{dN_2}{dt} = -(k_3 - k_4 N_1)N_2. \tag{10.1}$$

These equations should be solved under suitable initial conditions for populations of predator and prey. Three of the four constants in this system can be taken to be equal to 1 for the sake of simplicity by choosing corresponding units to measure the quantities N_1, N_2 and t. Thus, we obtain from (10.1) more simple system of equations

$$\frac{dN_1}{dt} = (1 - N_2)N_1, \tag{10.2}$$

$$\frac{dN_2}{dt} = -k(1 - N_1)N_2, \tag{10.3}$$

which contains only one parameter k. This parameter determines the relative degree of reproduction of predators with respect to prey.

Multiplying (10.2) by k and adding to (10.3) we find:

$$\frac{d}{dt}(kN_1 + N_2) = k(N_1 - N_2)$$

$$= k[(1 - N_2) - (1 - N_1)]$$

$$= k\left[\frac{1}{N_1}\frac{dN_1}{dt} + \frac{1}{kN_2}\frac{dN_2}{dt}\right].$$

Here we have use Eqs. (10.2 – 3) in the derivation of the last term. Simple integration gives the result

$$kN_1 + N_2 = k \ln N_1 + \ln N_2 + \text{const},$$

or

$$\frac{\exp(kN_1)}{N_1^k} = N_2 \exp(C - N_2). \tag{10.4}$$

This equation relates the populations of predators and prey with each other. The constant C is determined by the initial values of their populations.

Now we discuss the various analytical solutions of the nonlinear system (10.2 – 3).

10.2 ELLIPTICAL STATIONARY POINT

It is seen from Eqs. (10.2 – 3) that the simplest solution is

$$N_1(t) = N_2(t) = 1.$$

The populations do not change in this case, if the initial conditions coincide with this relation.

Let us find the solutions in the vicinity of this stationary point. We put

$$N_1 = 1 + \delta N_1; \qquad N_2 = 1 + \delta N_2; \qquad \delta N_1, \delta N_2 \ll 1.$$

Substituting these relations into (10.2 – 3) and linearizing the system we obtain the simple linear system of two differential equations

$$\frac{d\delta N_1}{dt} = -\delta N_2; \qquad\qquad \frac{d\delta N_2}{dt} = k\delta N_1. \qquad (10.5)$$

It follows from this system that

$$\frac{d^2 \delta N_1}{dt^2} + k\delta N_1 = 0.$$

This equation has the solution

$$\delta N_1(t) = a \cos(\sqrt{k}\, t + \varphi);$$

$$\delta N_2(t) = a\sqrt{k}\, \sin(\sqrt{k}\, t + \varphi). \qquad (10.6)$$

Here a and φ are the amplitude and phase of oscillations of the populations around the stationary point. The period of oscillations is equal to $T = 2\pi/K^{1/2}$. In the plane (N_1, N_2) the solution (10.6) corresponds to an ellipse (see Fig. 18), and its equation is

$$(\delta N_1)^2 + \frac{1}{k}(\delta N_2)^2 = a^2. \qquad (10.7)$$

Thus, the equilibrium point is stable; populations of prey and predators oscillate around this point with small amplitudes.

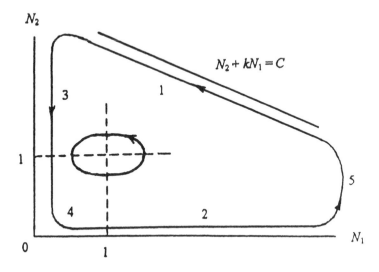

Figure 18. Dynamics of populations of predators (N_2) and prey (N_1) for different initial conditions.

10.3 ASYMPTOTIC SOLUTIONS OF THE PREDATOR – PREY PROBLEM

Let us find the solution of the system (10.2 – 3) for large populations of predators and prey, i.e. under the conditions $N_1 \gg 1$, $N_2 \gg 1$. Equation (10.2) can be simplified to

$$\frac{dN_1}{dt} = -N_1 N_2. \tag{10.8}$$

Simplifying analogously Eq. (10.3) and using (10.8), we find

$$\frac{dN_2}{dt} = kN_1 N_2 = -k \frac{dN_1}{dt}. \tag{10.9}$$

Thus, it follows from Eq. (10.9) that we have the simple solution

$$N_2 + kN_1 = C \gg 1. \tag{10.10}$$

This is depicted by straight line with index "1" in Fig. 18. Of course, this solution can be also obtained from Eq. (10.4) if we include exponents only at large values of populations. The time dependence of populations on this straight line can be found by substitution of (10.10) into (10.8):

$$\frac{dN_1}{dt} = -(C - kN_1)N_1. \tag{10.11}$$

Integrating this equation we obtain

$$N_1(t) = \frac{C}{k(1 + \exp(Ct))}; \qquad N_2(t) = \frac{C}{1 + \exp(-Ct)}. \tag{10.12}$$

The integration constant can be excluded by appropriate choice of the initial time.

It follows from Eq.(10.12) that the restrictions for populations of predators and prey (see Fig. 18) are:

$$N_1(t \to -\infty) = N_{1\,max} = \frac{C}{k};$$

$$\tag{10.13}$$

$$N_2(t \to \infty) = N_{2\,max} = C.$$

Thus, the coefficient k represents the ratio of the maximum number of predators to the maximum number of prey. The maximum number of predators corresponds to the minimum number of prey and *vice versa*. Oscillations of the populations always takes place; the direction of the oscillations is shown in Fig. 18.

The index "2" on Fig. 18 describes that part of the asymptotic curve which corresponds to the Malthus law for prey for a small population of predators. According to Eq. (10.2) we have in this case

$$N_1(t) = \frac{C}{k} \exp(t). \tag{10.14}$$

Here the time $t=0$ corresponds to the maximum number of prey. The number of predators on this part of the curve is, according to Eq. (10.4), given approximately by the relation

$$N_2 = \frac{\exp(kN_1 - C)}{N_1^k}.$$

The number of predators is minimal at $N_1 = 1$. It is small everywhere on part "2" of the curve since $C \gg 1$.

Let us now consider the vertical part "3" of this curve. Here predators die due to the absence of food (prey). According to Eq.(10.3) we have

$$N_2(t) = C \exp(-kt). \qquad (10.15)$$

Here the time $t = 0$ corresponds to the maximal value of predators (see Fig. 18). The population of prey is minimal at $N_2 = 1$ according to Eq. (10.4).

Region "4" near the origin corresponds to $N_1, N_2 \ll 1$. Then it follows from (10.4) that this part of curve is described by the dependence

$$N_2 N_1^k = \exp(-C) \ll 1. \qquad (10.16)$$

This part of the curve is similar to a hyperbola.

Analogously we can describe region "5" of the curve in the vicinity of the maximum number of prey. It follows from (10.4) that for $N_1 \gg 1$

$$\exp(kN_1) = N_2 \exp(C - N_2).$$

The maximum of the right side of this relation is achieved at $N_2 = 1$, and at this point $N_1 \approx C/k$. The arrows in Fig. 18 show the direction of time.

10.4 PREDATORS EATING EACH OTHER

The second part of this chapter is devoted to the problem of interacting populations of two types of predators which can eat each other (but with different voracity). Instead of (10.2 – 3) we have the system of equations for populations N_1, N_2 of these two types of predators (of course, predators do not eat the same type of predators according to biological laws):

$$\frac{dN_1}{dt} = -(1 - N_2)N_1; \qquad (10.17)$$

$$\frac{dN_2}{dt} = -k(1 - N_1)N_2. \qquad (10.18)$$

It is seen that the only difference is that the sign of the right side of Eq. (10.17) is opposite to the sign of the right side of Eq. (10.2). However this difference strongly influences the solution of the system. The parameter k determines the relative voracity of the predators 1 and 2. Now we can repeat the derivations used in obtaining of Eq. (10.4). Instead of this equation we find

$$\frac{N_2}{N_1^k} = \exp(N_2 - kN_1 - C). \qquad (10.19)$$

The constant C is determined by the initial populations of predators.

10.5 HYPERBOLIC STATIONARY POINT

It follows from (10.17 – 18) that $N_1(t) = N_2(t) = 1$ is again one of the solutions. Let us consider the vicinity of this stationary point. Putting

$$N_1 = 1 + \delta N_1, \qquad N_2 = 1 + \delta N_2; \qquad \delta N_1, \delta N_2 \ll 1$$

we obtain, instead of (10.5), the linear system of equations

$$\frac{d\delta N_1}{dt} = \delta N_2; \qquad \frac{d\delta N_2}{dt} = k\delta N_1. \qquad (10.20)$$

The simple solution of this system is of the form (instead of (10.6)):

$$\delta N_1(t) = a \cosh(\sqrt{k}t + \varphi);$$

$$\delta N_2(t) = a\sqrt{k} \sinh(\sqrt{k}t + \varphi). \qquad (10.21)$$

This is an unstable solution unlike (10.6). Instead of elliptical motion according to Eq. (10.7) we obtain the hyperbolic curves

$$(\delta N_1)^2 - \frac{1}{k}(\delta N_2)^2 = a^2. \qquad (10.22)$$

They are shown by dashed lines in Fig. 19. These lines are known as *separatrices*. Thus, the solutions around the stationary point are unstable, and there are no oscillations of populations in this case.

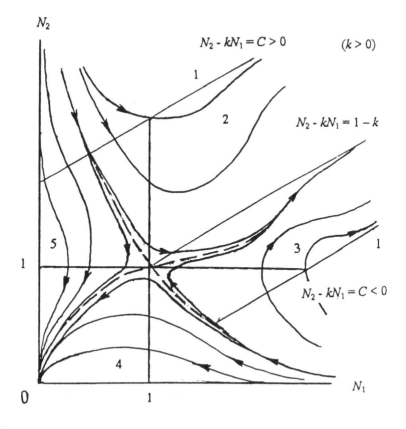

Figure 19. Dynamics of interacting populations of two types of predators

10.6 ASYMPTOTIC SOLUTIONS OF THE PREDATORS 1 – PREDATORS 2 PROBLEM

We consider here the asymptotic solution of (10.17 – 18) in the case when the populations of both types of predators are large, i.e. N_1, $N_2 \gg 1$. It follows from (10.19) that

$$N_2 - kN_1 = C . \qquad (10.23)$$

Unlike (10.10), now the constant C can be both large and small (and even negative). The two straight lines marked by "1" show this dependence in Fig. 19 in the cases $C > 0$, and $C < 0$, respectively. The asymptotic straight line of the separatrix lies between them and is described by the equation

$$N_2 - kN_1 = 1 - k .$$

Indeed, the value of the constant $C = 1 - k$ follows from Eq. (10.19) when $N_1 = N_2 = 1$. This separatrix is described by (10.22) with $a = 0$ in the vicinity of the stationary point, i.e. by the straight line $\delta N_2 = k\delta N_1$.

Now we consider solutions with $N_2 \gg 1$ and $C \gg 1$ marked by curve "2" in Fig. 19. The value of N_1 can be arbitrary. We have, from Eq. (10.19),

$$\exp(N_2) = \frac{1}{N_1^k} \exp(kN_1 + C). \qquad (10.24)$$

The minimum of N_2 is achieved at $N_1 = 1$, and the minimal value of N_2 is equal to C. Thus, in this regime first the number of predators 2 decreases, but then their number increases up to the asymptotic value given by (10.23).

Further we describe the solutions with $N_1 \gg 1$ and $-C \gg 1$ (this constant is now negative). We have from Eq. (10.19)

$$\exp(kN_1) = \frac{1}{N_2} \exp(N_2 - C). \qquad (10.25)$$

This is curve "3" in Fig. 19. The minimum of N_1 is achieved at $N_2 = 1$. The minimal value of N_1 is equal to $-C/k$.

Finally we consider solutions from the other side of the separatrix δN_2 = $k\delta N_1$. According to (10.19) this separatrix is of exponential decreasing form at $N_1 \gg 1$:

$$N_2 = N_1^k \exp(k - 1 - kN_1) \ll 1. \tag{10.26}$$

An analogous form is obtained on the other side of this separatrix.

Curve "4" in Fig. 19 is described by an equation which is also obtained from (10.19)

$$N_2 = N_1^k \exp(C - kN_1). \tag{10.27}$$

The population of predators 2 is growing at first, achieves a maximum at N_1 = 1 and then decreases. The population of predators 1 decreases instantaneously. Finally both types of predators disappear (at the origin).

The law of the disappearance of predators can be obtained again from (10.19) by putting $N_1, N_2 \ll 1$:

$$N_2 = \exp(-C)N_1^k. \tag{10.28}$$

The equation of the separatrix in this region is described by (10.28) with C = 1- k.

Equation (10.28) is valid also for curve "5" which also describes the disappearance of predators, but with a negative value of the constant C. The law of exponential decay follows from Eqs. (10.17 – 18) in the vicinity of the origin both for curve "4" and for curve "5":

$$N_1 \propto \exp(-t); \qquad N_2 \propto \exp(-kt). \tag{10.29}$$

We conclude that interacting populations of two types of predators are unstable. Depending on the initial populations, both populations can disappear, or increase to infinity simultaneously, so that $N_2 = kN_1 \to \infty$.

10.7 EPIDEMIC ILLNESS OF PREY

In the derivation of Eq. (10.5) we assumed that there are no restrictions in the increase of the population of preys, besides that produced by predators. However, prey can also die due to epidemic illness. We consider this problem in the vicinity of the elliptical stationary point (Section 10.2), for the sake of mathematical simplicity.

Instead of the first of the Eqs. (10.5) we can now write, for the deviation of the population of prey from its equilibrium value, $\delta N_1 = N_1 - 1$:

$$\frac{d\delta N_1}{dt} = -\delta N_2 - \alpha(1 + \delta N_1)^2 \approx -\delta N_2 - \alpha(1 + 2\delta N_1); \tag{10.30}$$

$$0 < \alpha \ll 1, \qquad \delta N_1 \ll 1, \qquad \delta N_2 \ll 1.$$

On the right side of this equation we have added a new term which represents a small decrease of the population of prey due to an epidemic illness.

Differentiating Eq. (10.30) and substituting the second of Eqs. (10.5), we obtain the differential equation for the small deviation δN_1 of the population of prey from the equilibrium value

$$\frac{d^2 \delta N_1}{dt^2} + k\delta N_1 = -2\alpha \frac{d\delta N_1}{dt}. \tag{10.31}$$

Its solution is ($a \ll 1$):

$$\delta N_1(t) = a \exp(-\alpha t) \cos(\sqrt{k}t + \varphi). \tag{10.32}$$

Hence,

$$\delta N_2(t) = a\sqrt{k} \exp(-\alpha t) \sin(\sqrt{k}t + \varphi) - \alpha. \tag{10.33}$$

Unlike the solution (10.6) we obtain the phase trajectory in the form of a spiral with the end at the equilibrium point $N_1 = 1$, $N_2 = 1 - \alpha \approx 1$.

Now we should take into account that predators do not like to eat ill or dead prey. Instead of the second of the Eqs. (10.5) we find

$$\delta N_2(t) = k\int \delta N_1(t)dt - \beta(\delta N_1)^3. \tag{10.34}$$

Then the additional number of predators decreases with increasing of prey ($\delta N_1 > 0$, more ill prey) and increases with decreasing of prey ($\delta N_1 < 0$).

Differentiating Eq. (10.30) and substituting Eq. (10.34), we obtain instead of Eq. (10.31):

$$(\delta N_1)'' + [2\alpha - 3\beta(\delta N_1)^2](\delta N_1)' + k(\delta N_1) = 0. \qquad (10.35)$$

This is the Van der Pol equation (see Eq. (4.46)). All phase trajectories end in the same limiting elliptical cycle (see Eq. (4.57)):

$$(\delta N_2)^2 + k(\delta N_1)^2 = 8\alpha k / 3\beta \ll 1. \qquad (10.36)$$

PROBLEMS

Problem 1. Consider the problem of two populations 1 and 2 which are fighting for common food. Obtain the system of nonlinear differential equations for these populations in the form

$$\frac{dN_1}{dt} = [1 - (N_1 + N_2)]N_1;$$

$$\frac{dN_2}{dt} = [k - k'(N_1 + N_2)]N_2.$$

Show that the relation between the populations is

$$N_2 = N_1^{k'} \exp[(k - k')t].$$

What are the conclusions about the dynamics of the populations with time depending on the parameters?

Chapter 11

Random Processes

This chapter is devoted to the mathematical analysis of physical processes which occur as a result of various random interactions in many-body systems. Instead of a general approach, we consider here some typical examples from different branches of theoretical physics.

11.1 AN ADDITIVE FUNCTION OF RANDOM VARIABLES

Let us consider an additive function of a large number of random variables:

$$S = \sum_m f(x_m).$$

For example, this is the potential acting on a chosen atom from surrounding atoms. Then x_m is the random coordinate of one of these atoms, and f is the potential of the mth atom. Then the quantity S is the resulting potential from all surrounding atoms. Of course, this problem can be both one-dimensional and multidimensional.

The instantaneous distribution of the values of S is described, obviously, by the Dirac delta-function:

$$F(S) = \delta\left(S - \sum_m f(x_m) \right)$$

Our goal is to average this distribution normalized by 1 over the random variables x_m.

177

We restrict ourselves here to the simplest case when all values of the random coordinate x_m have *the same probability* (the general case of different probabilities is considered analogously by means of the introduction of the corresponding probability function, see the next section). Let us denote the interval for averaging as $-L/2 < x_m < L/2$ $(L \to \infty)$. Then we obtain

$$< F(S) > \equiv \int_{-L/2}^{L/2} \frac{dx_1}{L} \int_{-L/2}^{L/2} \frac{dx_2}{L} \dots \int_{-L/2}^{L/2} \frac{dx_N}{L} \delta\left(S - \sum_m f(x_m)\right)$$

(11.1)

Here N is the number of random variables.

We use the known exponential representation of the Dirac delta-function:

$$\delta\left(S - \sum_m f(x_m)\right) = \frac{1}{2\pi} \int_{-\infty}^{\infty} \exp\left[i\left(S - \sum_m f(x_m)\right)t\right] dt.$$

(11.2)

Substituting (11.2) into (11.1), we obtain the averaged distribution function in the form

$$< F(S) > = \frac{1}{2\pi} \int_{-\infty}^{\infty} dt \exp(iSt) \left[\int_{-L/2}^{L/2} \frac{dx_m}{L} \exp(-if(x_m)t)\right]^N.$$

(11.3)

The square brackets in this expression can be identically rewritten in the form

$$[\dots] = \left[1 - \frac{(N/L)}{N} \int_{-L/2}^{L/2} dx\left(1 - \exp(-if(x)t)\right)\right]^N.$$

(11.4)

Here we omit the subscript m in the integrand variable. The quantity $n = N/L$ is the concentration of particles (in three-dimensional space it would be the number of particles per unit volume). We assume that $n = $ const, as $N, L \to \infty$. Thus, the concentration of particles is fixed.

Hence, Eq. (11.4) can be rewritten once more in the form

$$[...] = \lim_{N \to \infty} \left[1 - \frac{n}{N} \int_{-\infty}^{\infty} dx \left(1 - \exp(-i f(x)t) \right) \right]^N$$

$$= \exp \left\{ - n \int_{-\infty}^{\infty} dx \left(1 - \exp(-i f(x)t) \right) \right\}.$$

Here we have used the well-known Euler representation for the exponential function.

The final result for the averaged distribution function (11.3) is

$$< F(S) > = \frac{1}{2\pi} \int_{-\infty}^{\infty} dt \, \exp(iSt) \, \exp \left\{ - n \int_{-\infty}^{\infty} dx \left(1 - \exp(-i f(x)t) \right) \right\}.$$

$$(11.5)$$

This expression can be used for the solution of many physical problems.

Let us consider now the simple example: $f(x) = 1/x^2$ (Coulomb forces). Then the integral in (11.5) can be derived explicitly:

$$\int_{-\infty}^{\infty} dx \left(1 - \exp(it / x^2) \right) = 2\sqrt{\pi t} \, \exp(i\pi / 4).$$

Substituting this result in Eq. (11.5), we obtain finally

$$< F(S) > = n \sqrt{\frac{1}{S^3}} \, \exp \left(- \frac{\pi n^2}{S} \right) \qquad (11.6)$$

It is seen that $< F(S=0) > = < F(S=\infty) > = 0$. The quantity n^2 is the width of this distribution. The dependence (11.6) is shown in Fig. 20.

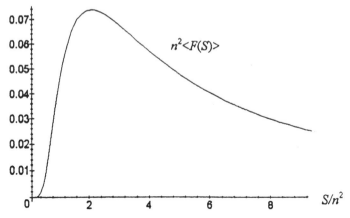

Figure 20. The distribution in Eq. (11.6).

The distribution function (11.6) has a maximum at

$$S = \frac{2\pi}{3} n^2 ; \qquad n^2 < F(S) >_{max} = \left(\frac{3}{2\pi e} \right)^{3/2} .$$

It follows from Eq. (11.6) that the normalization condition is:

$$\int_0^\infty < F(S) > dS = 1,$$

as it should be (since the initial distribution was normalized).

If $n^2 \ll 1$, then Eq. (11.6) reduces to the initial Dirac delta-function, as it should do, and is in agreement with the physical picture.

11.2 CENTRAL LIMIT THEOREM

In the previous section we assumed that all values of each random variable have *the same* probability. Now we consider the more general case when the probability of finding the value of the coordinate $x = x_m$ is equal to $w(x_m)$. For the sake of simplicity we consider the sum of these random variables:

$$S = \sum_{m=1}^{N} x_m. \qquad (11.7)$$

Here $N \to \infty$. We can put $<x_m> = 0$, since if this averaged value is nonzero, then we can consider the new random variable $y_m = x_m - <x_m>$.

We assume that the dispersions of the random variables are nonzero and the same for all variables:

$$<x_m^2> = \int_{-\infty}^{\infty} x^2 w(x)dx = \sigma. \qquad (11.8)$$

Here $w(x)$ is the probability distribution for each random variable. In order for the integral (11.8) to be convergent, this distribution should satisfy the condition

$$w(x) \xrightarrow[x \to \infty]{} \frac{1}{x^k}; \qquad k > 3. \qquad (11.9)$$

Our goal is to calculate the distribution function $W(S)$ for the sum (11.7). Let us introduce the so-called *characteristic function*

$$<\exp(iux)> = 1 - \frac{1}{2}u^2\sigma + ...$$

Then the characteristic function for the sum of independent random variables is of the form (we use the Euler representation for the exponential, analogously to the previous section):

$$<\exp(iuS)> = \left(1 - \frac{1}{2}u^2\sigma + ...\right)^N = \left(1 - \frac{1}{2N}Nu^2\sigma + ...\right)^N$$

$$= \exp\left(-\frac{1}{2}Nu^2\sigma\right) = \int_{-\infty}^{\infty} \exp(iuS)W(S)dS.$$

In order to find the distribution function of the sum (11.7), we use the inverse Fourier transformation in the last expression. Then we obtain

$$W(S) = \frac{1}{2\pi} \int_{-\infty}^{\infty} \exp[-iuS - Nu^2\sigma/2]du = \frac{1}{\sqrt{2\pi\sigma N}} \exp\left\{-\frac{S^2}{2\sigma N}\right\}.$$

This distribution depends on the number N of random variables. It is seen that if we wish to obtain a sum of random variables which does not depend on N, we should determine the quantity

$$s = \frac{1}{\sqrt{N}} \sum_{m=1}^{N} x_m.$$

Then the final expression for the distribution function is

$$w(s) = \frac{1}{\sqrt{2\pi\sigma}} \exp\left(-\frac{s^2}{2\sigma}\right) \qquad (11.10)$$

The most important property of this distribution is that it does not depend on the concrete form of the distribution function $w(x)$ of each random variable. This property is the so-called *central limit theorem*.

11.3 CAUCHY DISTRIBUTION

The results of the previous section are correct under the condition (11.9) for the distribution function of each random variable. Now we consider the case $k = 2$, so that (11.9) is not fulfilled. Again we put the averaged value of each random variable equal to zero. Let each random variable have a normalized probability distribution in the form (the *Lorentz distribution*)

$$w(x) = \frac{a}{\pi(a^2 + x^2)}.$$

We derive the characteristic function for this distribution (see the previous section):

$$< \exp(iux) > = \int_{-\infty}^{\infty} \exp(iux) \frac{adx}{\pi(a^2 + x^2)} = \exp(-a|u|).$$

Then for the sum (11.7) of independent random variables we obtain the characteristic function

$$< \exp(iuS) > = \exp(-Na|u|) = \int_{-\infty}^{\infty} \exp(iuS)W(S)dS.$$

Using again the inverse Fourier transformation, we find

$$W(S) = \frac{1}{2\pi} \int_{-\infty}^{\infty} \exp\left(- Na|u| - iuS \right)du = \frac{Na}{\pi(S^2 + N^2a^2)}.$$

It is seen that in order to obtain a distribution which doe not depend on the number N of random variables, we should determine, instead of (11.7), the quantity

$$s = \frac{1}{N} \sum_{m=1}^{N} x_m.$$

Then the distribution function for this quantity (the *Cauchy distribution*) is of the same form as the initial distribution function for each random variable

$$W(s) = \frac{a}{\pi(a^2 + s^2)}.$$

11.4 CAMPBELL'S THEOREM

The next well-known physical example is the derivation of the sum of additive random short pulses

$$S = \sum_{m=1}^{N} f(t - t_m). \tag{11.11}$$

Here t_m is the random time of the pulse. We use the simplest form of rectangular pulse:

$$f(t) = 1, \qquad 0 < t < \tau;$$

$$f(t) = 0, \qquad t < 0, \quad t > \tau.$$

The time T for averaging of sum is assumed to be large compared with τ.
First we derive the average value of one pulse

$$< f > = \frac{1}{T} \int_0^T f(t - t_m) dt_m = \frac{1}{T} \int_{t-T}^t f(\theta) d\theta \approx \frac{1}{T} \int_{-\infty}^{\infty} f(\theta) d\theta = \frac{\tau}{T} \ll 1.$$

Hence, the average value of the sum (11.11) is

$$< S > = N \frac{\tau}{T} = n\tau. \tag{11.12}$$

Here $n = N/T$ is the number of pulses per unit time. We assume that $n\tau \ll 1$, in order that the neighbouring pulses do not overlap each other.
Now we derive the quantity

$$< S^2 > = \sum_{m=1}^{N} < f^2(t - t_m) > + \sum_{m \neq j}^{N} \sum^{N} < f(t - t_m) >< f(t - t_j) >$$

$$= N \frac{\tau}{T} + N(N - 1)\left(\frac{\tau}{T}\right)^2 \approx \left(\frac{N\tau}{T}\right)^2 + \frac{N\tau}{T}.$$

Thus, the dispersion of the sum is

$$< \Delta S^2 > = < S^2 > - < S >^2 = n\tau. \tag{11.13}$$

The expressions (11.12 – 13) are the content of the so-called *Campbell theorem* for the sum of a large number of random short pulses. Of course, this theorem is correct for an arbitrary form of pulse. Then τ is the average time of each pulse.

11.5. CORRELATION FUNCTION FOR RANDOM PHASES

Let us consider a function $f(t)$. We can expand it in a Fourier series (or Fourier integral) and consider only one Fourier component of this function with frequency ω. Thus, we note (first without any phase perturbation): $f(t) = \exp(i\omega t)$.

We assume for the sake of illustrative simplicity that $t > 0$ is the time. Now we suggest that this function acquires the phase factor φ_m when the time is equal to *some random value* t_m. Hence, after a large number of such random processes our function takes the form

$$f(t) = \exp\left[i\omega t + i\Sigma_m \varphi_m \eta(t - t_m) \right]. \tag{11.14}$$

Here the notation

$$\eta(t) = \begin{cases} 0, & t < 0; \\ 1, & t > 0; \end{cases}$$

is introduced for the turn-on function (the Heaviside function). We now define *the correlation function* for the process ($t > 0$):

$$F(t) = < f^*(t')f(t'+t) >. \tag{11.15}$$

This correlation function contains the average over time $t' > 0$ during some large time interval. Our goal is to calculate this correlation function.

Let us further define the difference of the correlation functions:

$$\Delta F(t) = F(t) - \exp(-i\omega dt)F(t + dt). \qquad (11.16)$$

Here the small difference dt is assumed to be small compared to the time between the neighbouring phase jumps $t_m - t_{m-1}$. Hence, only one phase, or no phase can be acquired during the time dt.

Substituting (11.15) into (11.16), and taking into account (11.14), we obtain after simple algebraic derivations

$$\Delta F(t)$$

$$= \left\langle f^*(t')f(t'+t)\{1 - \exp[i\varphi_m(\eta(t'+t + dt - t_m) - \eta(t'+t - t_m))]\} \right\rangle.$$
$$(11.17)$$

Summing over m has been omitted here since the probability of double, triple and so on acquisitions of phase is very small.

Averaging over a small time interval dt in (11.17) can be carried out independent of averaging over other time intervals due to the random character of the process. The average value in Eq. (11.17) can thus be obtained as a product of individual averages, which gives:

$$\Delta F(t) = F(t)i\omega_0 dt, \qquad (11.18)$$

where the notation

$$i\omega_0 dt \equiv \left\langle 1 - \exp(-i\varphi_m) \right\rangle \Big|_{t'+t<t_m<t'+t+dt} \qquad (11.19)$$

is introduced. The quantity $\omega_0 = \text{Re}\,\omega_0 + i\,\text{Im}\,\omega_0$ is the specific complex constant of the system, depending on its details.

It follows from (11.18) and the Taylor expansion of (11.16) that

$$F(t)i\omega_0 dt = F(t)i\omega dt - dF(t).$$

This ordinary differential equation has the simple solution:

$$F(t) = \exp[i(\omega - \omega_0)t]. \qquad (11.20)$$

Here $t > 0$. Thus, we obtain the correlation function for the random process.

As an example, we consider the collision broadening of atomic spectral lines. Then φ_m is the scattering phase in the wave function $f(t)$. Averaging Eq. (11.19) over time amounts to averaging the collision times t_m contained within some large time interval. From the classical point of view, the impact parameter ρ of the collision is unambiguously connected with t_m, so it is convenient to carry out the averaging in terms of ρ. The collision probability per unit time in an interval of impact parameters between ρ and $\rho + d\rho$ is equal to $2\pi\rho d\rho Nv$, where N is the number density of the perturbing particles, and v is the relative velocity of the collision. Hence, the average required for Eq. (11.12) is

$$\left\langle 1 - \exp[i\varphi(\rho)]\right\rangle = dt \cdot vN \int_{0}^{\infty} 2\pi\rho d\rho[1 - \exp(i\varphi(\rho))].$$

We introduce the notation

$$\sigma_t = \int_{0}^{\infty} 4\pi\rho d\rho(1 - \cos\varphi)$$

for the total cross-section for elastic scattering, and

$$\Delta\omega = -Nv \int_{0}^{\infty} 2\pi\rho d\rho \cdot \sin\varphi$$

for the Stark shift of the atomic transition frequency. Then it follows from (11.20) that the correlation function is of the form

$$F(t) = \exp[i(\omega - \omega_0)t - Nv\sigma_t t / 2]. \tag{11.21}$$

The probability of emitting a photon with the frequency Ω is

$$w(\Omega) = \left| \frac{1}{\sqrt{2\pi}} \int \exp(-i\Omega t) f(t) dt \right|^2.$$

This integral can be written in the form

$$w(\Omega) = \frac{1}{2\pi} \int dt_1 \int dt_2 f(t_1) f^*(t_2) \exp[i\Omega(t_2 - t_1)].$$

In terms of the new variables $t' = t_2$, $t = t_1 - t_2$ we can rewrite this expression as (using the definition of the correlation function in Eq. (11.15))

$$w(\Omega) = \frac{1}{2\pi} \int_0^\infty F(t) \exp(-i\Omega t) dt + \text{c.c.} \tag{11.22}$$

Substituting (11.21) into (11.22), we find

$$w(\Omega) = \frac{1}{2\pi} \cdot \frac{Nv\sigma_t}{(\Omega - \omega + \Delta\omega)^2 + (Nv\sigma_t / 2)^2}. \tag{11.23}$$

We have found that the collision broadening of spectral lines exhibits a Lorentz form. We have both a broadening of the spectral line and a significant shift in its central frequency.

11.6 RANDOM MOTION OF A CLASSICAL PARTICLE

Here we consider the classical Newtonian motion of a particle induced by some random force $F(t)$ in a medium with friction. The one-dimensional Newton equation for the velocity $v(t)$ of a particle is of the form

$$\frac{dv}{dt} = -kv + \frac{1}{m} F(t). \tag{11.24}$$

Here m is the mass of the particle, and k is the coefficient of friction due to collisions with other particles. We assume that $v(0) = 0$ (it is an inessential condition). The simple solution of the differential equation (11.24) is

$$v(t) = \frac{1}{m} \int_0^t \exp[-k(t - t')]F(t') dt'. \tag{11.25}$$

We have for the averaged value of the random force $<F(t)> = 0$. It follows from Eq. (11.25) that $<v(t)> = 0$.

Let us now calculate the mean square value of the velocity:

$$< v^2(t) > = \frac{1}{m^2} \int_0^t dt' \int_0^t dt'' \exp[-2kt + k(t'+t'')] < F(t)F(t') > .$$

(11.26)

Thus, we obtain the correlation function $<F(t)F(t')>$ for the random force, similar to the correlation function in the previous section. We assume that the time for the correlation Δt (this is the duration of one force shock, for example) is very small compared both to the time $1/k$ for relaxation of a particle due to friction (weak friction) and to the time interval between the subsequent shocks. Then we can approximate

$$< F(t')F(t'') > = F^2 \Delta t \cdot \delta(t'-t'').$$

(11.27)

Here F is the amplitude value of the force. Substituting (11.27) into (11.26), we calculate the simple integral and obtain

$$< v^2(t) > = \frac{F^2 \Delta t}{2m^2 k} [1 - \exp(-2kt)].$$

(11.28)

If $kt \ll 1$, then it follows from (11.28) that

$$< v^2(t) > = 2D_v t; \qquad D_v = \frac{F^2 \Delta t}{2m^2}.$$

(11.29)

The quantity D_v is called the *diffusion coefficient*. It is seen that D_v does not depend on the coefficient of friction. In our case the diffusion coefficient is constant, i.e. it does not depend on the velocity of the particle either.

Conversely, if $kt \gg 1$, then the velocity takes the stationary value

$$< v^2(\infty) > = \frac{F^2 \Delta t}{2m^2 k}.$$

Let us introduce the distribution function $f(v,t)$ which determines the number of particles with velocity v for the time moment t. If the random force is absent, the velocities of all particles decrease as $v_0 \exp(-kt)$. Hence, the distribution function f is of the form

$$f(v, t) = f_0(v_0) \exp(kt) = f_0(v \exp(kt)) \exp(kt), \qquad (11.30)$$

where $f_0(v_0)$ is the distribution function for the initial time moment $t = 0$. Indeed, we find for the mean value of the velocity:

$$< v > = \int vf dv = \int v \exp(kt) f_0(v \exp(kt)) dv$$

$$= \int v_0 f_0(v_0) dv_0 \cdot \exp(-kt) = < v_0 > \exp(-kt).$$

as it should be. The equation for the distribution function is of the form

$$\frac{\partial f}{\partial t} = k \frac{\partial}{\partial v}(vf) = kf + kv \frac{\partial f}{\partial v}. \qquad (11.31)$$

This can be checked by direct substitution of Eq. (11.30) into Eq. (11.31).
    Further we omit friction, but introduce a random force. Then we can write

$$f(v + dv, t) = f(v, t) + \frac{\partial f}{\partial v} dv + \frac{1}{2} \frac{\partial^2 f}{\partial v^2} (dt)^2.$$

According to (11.29)

$$< dv^2 > = 2D_v dt.$$

Besides this, $< dv > = 0$. Hence, after stochastic averaging we obtain

$$f(v + dv) = f(v, t) + D_v dt \cdot \frac{\partial^2 f}{\partial v^2}.$$

The left side of this equation can be also written as $f(v, t+dt)$ since the total differential of the distribution function is equal to zero according to Liouville's theorem: the distribution function is constant along the phase trajectory. Hence,

$$\frac{\partial f}{\partial t} = D_v \frac{\partial^2 f}{\partial v^2}. \qquad (11.32)$$

Together with Eq. (11.31) this equation can be generalized as

$$\frac{\partial f}{\partial t} = k \frac{\partial}{\partial v}(vf) + D_v \frac{\partial^2 f}{\partial v^2}. \qquad (11.33)$$

This is the so-called *Fokker-Planck equation* (see also Section 9.4).

The stationary solution of this equation at large times (compared to $1/k$) is of the form

$$f = \text{const} \cdot \exp\left(-\frac{kv^2}{2D_v}\right) \qquad (11.34)$$

For example, in the case of a molecular gas we obtain the well-known *Maxwell equilibrium distribution* with diffusion coefficient $D_v = kT/m$. The last expression is the so-called *Einstein relation*. It should be noted that the Maxwell distribution is applicable not only for an ideal gas, but also for arbitrarily strong interactions between molecules.

Thus, we have derived the Fokker – Planck equation from the Newton equation for particles perturbed by mutual random shocks. It should be noted once more that it is not necessary for the system to be ideal (for example, an ideal molecular gas).

11.7 FLUCTUATIONS OF RANDOM QUANTITIES

This section is devoted to derivations of fluctuations of random variables. Instead of a general approach, we consider, as usual, a typical example: N noninteracting particles are found inside the volume V. Let us calculate the fluctuation of the number of particles n inside the smaller volume $v < V$ with respect to its obvious mean value $<n> = N(v/V)$.

The microscopic concentration of particles is expressed, obviously, via a sum of Dirac delta-functions:

$$\rho(\mathbf{r}) = \sum_{i=1}^{N} \delta(\mathbf{r} - \mathbf{r}_i).$$

The deviation of this real concentration from its mean value $< \rho(\mathbf{r}) > = N/V$ (since $< \delta(\mathbf{r} - \mathbf{r}_i) > = 1/V$) is

$$\delta\rho(\mathbf{r}) = \sum_{i=1}^{N} \delta(\mathbf{r} - \mathbf{r}_i) - \frac{N}{V}.$$

We now derive the correlation function for the concentration of particles:

$$< \delta(\mathbf{r})\delta(\mathbf{r}') > = < \sum_{i,j=1}^{N} \delta(\mathbf{r} - \mathbf{r}_i)\delta(\mathbf{r}'-\mathbf{r}_j) > - \left(\frac{N}{V}\right)^2. \qquad (11.35)$$

The sum on the right side of this expression can be written as

$$< \sum_{i=j}^{N} \delta(\mathbf{r} - \mathbf{r}_i)\delta(\mathbf{r}'-\mathbf{r}_j) > + < \sum_{i \neq j}^{N} \delta(\mathbf{r} - \mathbf{r}_i)\delta(\mathbf{r}'-\mathbf{r}_j) > . \qquad (11.36)$$

The second term in Eq. (11.36) is derived using the independence of motion of different particles:

$$< \sum_{i \neq j}^{N} < \delta(\mathbf{r} - \mathbf{r}_i)\delta(\mathbf{r}'-\mathbf{r}_j) >$$

$$= \sum_{i \neq j}^{N} < \delta(\mathbf{r} - \mathbf{r}_i) >< \delta(\mathbf{r}'-\mathbf{r}_j) > = \frac{N(N-1)}{V^2}.$$

The first term in Eq. (11.36) is derived directly:

$$< \sum_{i=j}^{N} \delta(\mathbf{r} - \mathbf{r}_i)\delta(\mathbf{r}'-\mathbf{r}_j) >$$

$$= \delta(\mathbf{r} - \mathbf{r}') \sum_{i=1}^{N} < \delta(\mathbf{r} - \mathbf{r}_i) > = \frac{N}{V}\delta(\mathbf{r} - \mathbf{r}').$$

Substituting these results into Eq. (11.35), we obtain the final simple expression for the correlation function

$$< \delta\rho(\mathbf{r})\delta\rho(\mathbf{r}') > = \frac{N}{V}\delta(\mathbf{r} - \mathbf{r}') - \frac{N}{V^2}. \qquad (11.37)$$

Double integration of Eq. (11.37) over the volume v results in the final expression for the dispersion of the number of particles

$$< \delta n^2 > = \frac{N}{V}v - \frac{N}{V^2}v^2 = < n > \left(1 - \frac{v}{V}\right) \qquad (11.38)$$

The fluctuation is by definition the square root of the dispersion.
    If $v = V$, then, obviously, no fluctuations of the total number of particles take place. Conversely, if $v \ll V$, then it follows from Eq. (11.31) that

$$< \delta n^2 > = < n >. \qquad (11.39)$$

Of course, the general result (11.38) can be also obtained, using the binomial distribution for the number of particles.
    It should be noted that in the above consideration the condition $N \gg 1$ is not necessary. Therefore we can investigate, in particular, the case $N = 1$. It follows from Eq. (11.38) that then the dispersion is

$$< \delta n^2 > = \frac{v}{V}\left(1 - \frac{v}{V}\right)$$

The maximum value of the fluctuation is achieved at $v = V/2$ and is equal to the square root of the dispersion, i.e., to $1/2$. Thus, the value of the fluctuation is equal to the mean value of the number of particles inside the half-volume ($=1/2$), i.e. the fluctuation is very large.

In statistical physics we have $N \gg n \gg 1$, so that the fluctuation of the number of particles in a volume

$$\sqrt{< \delta n^2 >} = \sqrt{< n >}$$

is small compared to the mean number of particles $< n >$.

11.8 RANDOM PHASE APPROXIMATION

Let us obtain the dynamical equation for the fluctuations of a physical quantity. Analogously to the previous section, we choose, as an example, the concentration of particles. The instantaneous value of this concentration is

$$\rho(\mathbf{r}) = \sum_{i=1}^{N} \delta(\mathbf{r} - \mathbf{r}_i). \tag{11.40}$$

Its Fourier component is

$$\rho_{\mathbf{k}} = \int \rho(\mathbf{r}) \exp(-i\mathbf{k}\mathbf{r})d\mathbf{r} = \sum_{i=1}^{N} \exp(-i\mathbf{k}\mathbf{r}_i).$$

The first derivative of this component is equal to

$$\frac{d\rho_{\mathbf{k}}}{dt} = -i \sum_{i=1}^{N} (\mathbf{k}\mathbf{v}_i) \exp(-i\mathbf{k}\mathbf{r}_i).$$

Here $\mathbf{v}_i$ is the velocity of the ith particle. Analogously we derive the second derivative:

$$\frac{d^2\rho_{\mathbf{k}}}{dt^2} = \sum_{i=1}^{N} \left\{ -i\left(\mathbf{k}\frac{d\mathbf{v}_i}{dt}\right) - (\mathbf{k}\mathbf{v}_i)^2 \right\} \exp(-i\mathbf{k}\mathbf{r}_i). \tag{11.41}$$

According to Newton's equation of motion for particles interacting with each other by the potential $U(\mathbf{r}_i - \mathbf{r}_j)$ we have

$$M \frac{d\mathbf{v}_i}{dt} = - \sum_{j \neq i} \nabla U(\mathbf{r}_i - \mathbf{r}_j). \qquad (11.42)$$

Here M is the mass of the particle. We expand the interaction potential in a Fourier series (or Fourier integral):

$$U(\mathbf{r}) = \frac{1}{V} \sum_{\mathbf{k'}} U_{\mathbf{k'}} \exp(i\mathbf{k'}\,\mathbf{r}),$$

where V is the volume of the system. Hence,

$$\nabla U(\mathbf{r}) = \frac{i}{V} \sum_{\mathbf{k'}} \mathbf{k'} U_{\mathbf{k'}} \exp(i\mathbf{k'}\,\mathbf{r}).$$

Substituting this expression into Eq. (11.42), we find

$$\frac{d\mathbf{v}_i}{dt} = - \frac{i}{MV} \sum_{\mathbf{k'}} \sum_{j \neq i} \mathbf{k'} U_{\mathbf{k'}} \exp[i\mathbf{k'}(\mathbf{r}_i - \mathbf{r}_j).$$

Substituting this expression into the first term in the right side of Eq. (11.41), we obtain this term in the form

$$I_1 = - \frac{1}{MV} \sum_{\mathbf{k'}} (\mathbf{k}\mathbf{k'}) U_{\mathbf{k'}} \sum_{i=1}^{N} \exp[i(\mathbf{k'}-\mathbf{k})\mathbf{r}_i] \sum_{j \neq i} \exp(-i\mathbf{k'}\,\mathbf{r}_j). \qquad (11.43)$$

Now we use the fact that the variables $\mathbf{r}_i$ are *random* quantities (this is valid in gases, or liquids, but not in solids). Then we put $\mathbf{k'} = \mathbf{k}$ in (11.43) and rewrite this expression in the simpler form (this is the so-called *random phase approximation*):

$$I_1 = -\frac{N}{MV}k^2 U_{\mathbf{k}} \sum_{j \neq i} \exp(-i\mathbf{k}\mathbf{r}_j) = -\frac{n}{M}k^2 U_{\mathbf{k}}\rho_{\mathbf{k}}. \qquad (11.44)$$

Here $n = N/V$ is the macroscopic mean concentration of the particles. The condition $j \neq i$ is inessential because of the large number of particles, $N \gg 1$.

Next we consider the second term on the right side of Eq. (11.42). Due to the random orientation of the velocities of the particles we obtain

$$I_2 = -\sum_{i=1}^{N}(\mathbf{k}\mathbf{v}_i)^2 \exp(-i\mathbf{k}\mathbf{r}_i) = -\frac{1}{3}k^2 \sum_{i=1}^{N}v_i^2 \exp(-i\mathbf{k}\mathbf{r}_i).$$

We have used here the mean value $< \cos^2\theta > = 1/3$ for the angle between the velocity of the particle and the wave vector.

The mean square of the velocity is the constant v^2 which does not depend on the index i of the particle. Hence, the last expression can be rewritten in the simple form

$$I_2 = -\frac{1}{3}(kv)^2 \rho_{\mathbf{k}}. \qquad (11.45)$$

Substituting Eqs. (11.44) and (11.45) into Eq. (11.41), we obtain the harmonic oscillator equation for the Fourier component of the concentration of particles:

$$\frac{d^2\rho_{\mathbf{k}}}{dt^2} + \omega_{\mathbf{k}}^2 \rho_{\mathbf{k}} = 0, \qquad (11.46)$$

where the notation

$$\omega_{\mathbf{k}}^2 = k^2 \left(\frac{1}{3}v^2 + \frac{n}{M}U_{\mathbf{k}} \right) \qquad (11.47)$$

is introduced.

It is seen from Eq. (11.46) that the small fluctuations of concentration of particles oscillate with time. In the limit of large wavelengths (i.e. small wave numbers k) it follows from Eq. (11.47) that the linear law for longitudinal sound waves is

$$\omega_{\mathbf{k}} = ck; \qquad c = \sqrt{\frac{1}{3}v^2 + \frac{n}{M}V_0}.$$ (11.48)

The quantity c can be considered as the speed of sound. It depends on the interaction between the particles. The first term in the speed of sound corresponds to an ideal gas of particles. It is also seen that the speed of sound is of the order of the mean velocity of particles v.

In the general case, the phase velocity of the waves depends on the frequency according to Eq. (11.47), i.e. dispersion of the waves takes place.

PROBLEMS

Problem 1. The electric field strength from Coulomb centers with charge q each of which are chaotically distributed in space (for example, atomic ions in an ideal gas) is determined by the Coulomb law

$$\mathbf{E} = q\sum_i \frac{\mathbf{r}_i}{r_i^3}.$$

Show that the normalized averaged distribution for values of this strength is of the form (the so-called *Holzmark distribution*)

$$< F(E) >= \frac{2}{\pi E} \int_0^\infty x \sin x \cdot \exp\left[-\left(\frac{xE_0}{E}\right)^{3/2} \right] dx;$$

$$E_0 \equiv 2\pi q \left(\frac{4n}{15}\right)^{2/3}.$$

Here n is the fixed concentration of Coulomb centers. Calculate the integral in the limiting cases (1) $E \gg E_0$, and (2) $E \ll E_0$.

Problem 2. Show that the central limit theorem is also valid for different probability distributions $w_m(x_m)$ of random variables. It is required only that all of these distributions satisfy the condition (11.9).

REFERENCES

1. Lichtenberg, A.J. and Liberman, M.A., *Regular and Stochastic Motion* (Springer, New York, 1983).
2. Landau, L.D. and Lifshitz, E.M., *Quantum Mechanics: Non-Relativistic Theory,* 3nd edition (Pergamon, Oxford, 1977).
3. Feynman, R.P. and Hibbs, A.R., *Quantum Mechanics and Path Integrals* (McGraw-Hill, New York, 1965).
4. *Handbook of Mathematical Functions (with formulas, graphs and mathematical tables),* ed. by M. Abramowitz and I.A. Stegun (Nat. Bureau of Standards, New York, 1964).
5. Mathews, J. and Walker, R.L., *Mathematical Methods in Physics* (Benjamin, New York, 1964).
6. Mott, N.F. and Massey, H.S.W., *The Theory of Atomic Collisions* 3nd edition (Clarendon Press, Oxford, 1965).
7. Gibbs, K., *Advanced Physics* (Cambridge University Press, Cambridge, 1988).
8. Nayfeh, A.H., *Introduction to Perturbation Techniques* (Wiley, New York, 1981).
9. Moon, F.C., *Chaotic Vibrations* (Wiley, New York, 1988).
10. March, N.H., Young, W.H. and Sampanthar, S. *The Many-Body Problem in Quantum Mechanics* (Cambridge University Press, Cambridge, 1967).
11. Migdal, A.B., and Krainov, V.P., *Approximation Methods in Quantum Mechanics* (Benjamin, New York, 1969).
12. Lotka, A.J., *Elements of Physical Biology* (Dover, New York, 1956).
13. Nicolis, G. and Prigogine, I., *Self-Organization in Nonequilibrium Systems* (Wiley, New York, 1977).
14. Cohen-Tannoudji, C., *Atom – Photon Interactions* (Wiley, New York, 1992).
15. Olver, F.W.J., *Asymptotics and Special Functions* (Academic Press, New York, 1974).
16. Dingle, R.B., *Asymptotic Expansions: Their Derivation and Interpretation* (Academic Press, London, 1973).
17. Watson, G.N., *A Treatise on the Theory of Bessel Functions* (Cambridge University Press, Cambridge, 1958).
18. Whittaker, E.T. and Watson, G.N., *A Course of Modern Analysis* (Cambridge University Press, Cambridge, 1952).
19. Courant, R. and Hilbert D., *Methods of Mathematical Physics*, Vol. 1 (Interscience Publishers, New York, 1953).
20. Feller, W., *Probability Theory and its Applications* (Wiley, New York, 1957).

Index